T0121906

BROADBANDITS

Inside the $750 Billion Telecom Heist

OM MALIK

WILEY

JOHN WILEY & SONS, INC.

Published by John Wiley & Sons, Inc., Hoboken, New Jersey.
Published simultaneously in Canada.

For general information on our other products and services, or technical support, please contact our Customer Care Department within the United States at 800-762-2974, outside the United States at 317-572-3993 or fax 317-572-4002.

Wiley also publishes its books in a variety of electronic formats. Some content that appears in print may not be available in electronic books.

For more information about Wiley products, visit our web site at www.wiley.com.

Library of Congress Cataloging-in-Publication Data:
Malik, Om.
    Broadbandits : inside the $750 billion telecom heist / Om Malik.
        p. cm.
Includes index.
    ISBN 0-471-43405-1 (cloth)
    1. Telecommunication—Corrupt practices—United States. 2. Consolidation and merger of corporations—United States. 3. Business failures—United States. I. Title.
    HE7775 .M345 2003
    384' .0973—dc21
                                                                        2002156438

10   9   8   7   6   5   4   3   2   1

For my grandfather and my parents

CONTENTS

BROADBANDITS: THE MOST WANTED LIST

Broadband (noun)   High-speed network access.

Broadbandit (noun)   One who padded his coffers by $50 million or more riding the bandwidth bubble.

The Bosses

Gary Winnick, cofounder of Global Crossing, who holds the record for becoming a billionaire faster than anyone in U.S. history. Meanwhile, his company blew through $15 billion in investors' money in less than five years and has become the fourth largest bankruptcy case in U.S. history. His take from Global Crossing: $735 million.

Bernie Ebbers, the WorldCom chief executive who used the broadband hype to boost the WorldCom stock, and then built a private fortune that includes ranches in Canada and timberlands in the American South.

Jack Grubman, known as telecom's pied piper—a telecom analyst at Salomon Smith Barney. During the height of the mania in 1999 and 2000, he had buy recommendations on 20 telecommunications companies. Twelve are now bankrupt and others are on the verge. For his efforts, Jack pocketed a cool $100 million.

Joe Nacchio, a former AT&T executive, who got his big chance when he became chief executive of Qwest Communications. He cleared around a quarter of a billion dollars in the five years he was at Qwest, while the company's market value plummeted. He made bad decisions, short-changed employees, ran the company into the ground, and is said to have allowed the fiber capacity swap scheme.

Ken Rice, the former chief executive of Enron Broadband Services and a former gas pipeline executive, who was more interested in his collection of Ferraris and motorcycles than in running Enron Broadband. For this, he made $70 million.

The Underbosses

Scott Sullivan, WorldCom's chief financial officer, who massaged numbers at WorldCom just so the stock wouldn't crash and founder Bernie Ebbers wouldn't go to jail.

Matt Bross, chief technology officer of Williams Communications, who pioneered a new kind of financial trick: the equity-for-customer-contract deal. Start-ups such as Corvis, ONI Systems, and Sycamore Networks paid Bross millions in stock to use their equipment.

Richard McGinn, the former CEO of Lucent, who was fired in October 2001 but received a severance package worth $12.5 million. Incompetence paid off handsomely—his total take for reducing Lucent to a shambles: about $38 million.

See the insider sales gang at the end of the book (Appendix B).

PROLOGUE

Inside the $750 Billion Telecom Heist

Poof—$750 billion gone! With over 100 companies bankrupt and an equal number that have shut shop, as many as 600,000 telecom workers are now without a paycheck. WorldCom is bankrupt, Global Crossing is decimated, PSINet has been sold for peanuts, and Genuity, a company as old as the Internet, sold its assets for a mere $250 million, a fraction of its one-time worth. These are staggering numbers for an industry that accounts for a sixth of the U.S. economy.

But they aren't as staggering as the amounts of money that hardworking employees at these companies have lost. After 31 years of relentless work, Lenette Crumpler, a former employee of Rochester, New York–based Frontier Communications, is without a job. As much as $86,000 of her 401(k) money went up in flames, through no fault of her own. Gary Winnick's Global Crossing bought Frontier and ruined most Frontier employees' savings like Crumpler's savings. Or take the case of Paula Smith, who worked most of her life at US West and then lost her entire life's savings of $400,000 after Qwest took over US West and, by extension, her 401(k) retirement plan. How will Kelsey and Ali, her daughters, go to college? And what about the 50-plus-year-old former telecom engineers who are now working part-time at Home Depot selling drills?

Even as they were cashing out their own holdings, the executives at these broadband companies encouraged employees to put their 401(k) dollars into company stock. And since the employees had nary a clue about these shenanigans, they complied and are now ruined. Now Crumpler, Smith, and the almost 600,000 other people in their situation face the figurative dole, while the robber barons of the information age sit in their Florida mansions, their Bel-Air palaces, and enjoy life on board their multimillion-dollar yachts. The robber barons of a century ago at least created industries of lasting value. But the modern-day robber barons used bankruptcy protection laws, lined their pockets, and walked into the sunset—just like two-bit bandits.

The biggest bubble in the history of the modern world was not the dot-com bubble but the telecom bubble. Sure, the Internet boom, and the ensuing bust, was more visible. People actually saw the web sites that sold things like drugs, pet food, and groceries collapse. But the broadbandits were another story. They were shortsighted, greedy, and their financial mismanagement left the telecommunications industry in shambles. Former phone company salesmen, stock market analysts, accidental entrepreneurs, greedy financiers, and executives who made out like bandits are the villains of this drama. Their antics would make Alexander Graham Bell roll over in his grave.

In stark contrast to the dot-com bust and the implosion of Enron, which unraveled with alarming speed, the disaster in the telecommunications industry arrived stealthily. What seemed like an endless demand for bigger and faster networks created a buildup of excessive proportions and a glut of capacity, the result being 600,000 jobs eventually evaporating into thin air. All this from an industry that at one point had a value of $2 trillion!

Beyond the quantity of money lost, the broadbandits' follies had a far broader effect on everyday Americans' lives as well as the country's financial elite. Visionary entrepreneurs lost their reputations; average joes lost their savings; investment banks became highly suspect; and the venture capital community, once the bulwark of free enterprise, became its own nemesis. The worst of it was that some essentially good ideas bit the dust. It was a case of the right thing at the right time with all the wrong people.

But while the broadbandits, bankers, and venture financiers deserve the most blame, it seems as if the entire sentient world acted in tandem—wittingly and unwittingly—to play out this Dickensian drama.

The broadband bubble, as I generally call it, had three aspects to it: the infrastructure bubble (Part I), the services bubble (Part II), and the equipment bubble (Part III).

It's Deja-Vu All Over Again

Telecom is not the first well-established industry to go through a boom and bust. The telegraph industry in the 1850s, the railroad market in the 1880s, the telephony sector in the 1890s, and the airline industry of the

1960s went through similar cycles. Most of these bubbles began with a technological innovation. Daniel Gross, writing in the *Milken Institute Review*, said that most of these infrastructure bubbles began when "entrepreneurs aimed first to link the business centers of the Northeast."[1] Capital was cheap, the economy was sound, and the stock market was booming. Significantly, most of these bubbles were marked by incompatible standards.

For example, when Samuel Morse invented the telegraph in 1842, he was able to raise cheap capital quickly, and by 1844 had set up a link between Washington and Baltimore. He soon extended the reach of his company, Magnetic Telegraph, to Jersey City, but by the end of the 1840s he was facing competition from the newbies. By 1852, the number of telegraph miles in the country rose to 23,000,[2] with another 10,000 in construction. The business was so competitive that price wars broke out, and even though telegraph technology became part of American life, only one company—Western Union—survived.

The railroad boom (1873–1894) was plagued by the same problems— price wars and the lack of universal standards (the railroad tracks weren't all the same width). When the dust settled, there were 224,000 railroad track miles, and seven groups—including J. P. Morgan and the Vanderbilts—owned them.[3] The telephone industry went through the same ups and downs between 1894 and 1913, with the initial construction of the disparate networks localized in the northeastern United States. A lot of companies went belly-up, but cheap prices brought in loads of new customers.

The broadband bubble is following the same pattern, despite the lessons history has taught us time and again. This time around, it proved to be a very expensive lesson. Like the railroad tracks that were built, with dubious logic, for travel to sparsely populated American towns, the world is crisscrossed with fiber that is unlikely to be used for decades. Unlit fiber networks snake through the American mainland and heartland, and cables as thick as a full-grown python lie dormant across the oceans.

According to KMI Research, about 80.2 million miles of optical fiber was installed in the United States during the six years between January 1, 1996, and December 31, 2001. That is almost 76 percent of the cumulative installed base of 105 million miles, based on KMI data going back to

1980.[4] TeleGeography, a consultancy that has mapped the networks of the world, estimates that as a result of the fiber-optic construction boom at the end of the twentieth century, most U.S. cities are now traversed by a thousand pairs of fiber, each one of them capable of carrying the entire phone traffic of the North American continent.

It's the same in Europe, where there is so much fiber under the streets of Geneva and London that it could give the entire continent bandwidth heartburn. TeleGeography research shows that the city of London, which is something of an international hub for bandwidth, has a capacity of 6.5 terabits per second, which is almost a million times the capacity of a 56 kilobits per second dial-up modem—enough to carry four times the amount of data and voice traffic currently emanating from 40 of the large European cities. Because of this surplus of bandwidth, circuits that used to cost $2 million now go for less than $200,000.

A Brave New World

In 1996, telecommunication was the new Wild West, thanks to a booming stock market and a series of coincidences.

The Telecommunication Act of 1996 had just been approved. It was a million-word-long piece of legislation that promised to unleash competition in the once-closed telephone industry and helped create new AT&Ts. This would result in an excess of bandwidth, where high-speed Internet access would be the norm, not an anomaly. The Federal Communications Commission (FCC) wanted to break the chokehold of the Baby Bells on the local market. Deregulation provided new opportunities for entrepreneurs—and so did the Internet.

After the novelty of e-mail and online services like America Online and CompuServe made headlines in 1994, the Internet became big news. In August 1995, the nascent Netscape Communications, which had turned the Mosaic Web browser from an esoteric university project to a must-have software download, had a wildly successful initial public offering (IPO), proving to the world the Internet's commercial possibilities. Internet service providers (ISPs) like Netcom, Earthlink, and PSINet were seeing exponential growth in their subscriber base, having started from zero. America Online, an upstart that quickly became the country's leading Internet service provider, saw its subscribers grow from 1.5 million in December 1994 to 7 million by December 1996, and to 15 million by December 1998.

These ISPs turned to UUNet and other larger companies to provide them with the bandwidth consumers were clamoring for. UUNet grew quickly between 1995 and 1996 and sold itself to MFS Communications, which in turn sold out to WorldCom, a second-tier long distance carrier WorldCom said its Internet business was booming, and that prompted Qwest, Global Crossing, Level 3, and other clones to get into the act. It all added up to fiber frenzy.

As luck would have it, the capital markets caught on to this new technology revolution and threw money at companies trying to conquer the newly open telecom market. Any stocks related to the Internet, broadband, and telecom skyrocketed. In short, money was the cheapest commodity in the world.

From 1996 to 1998, the bandwidth demand increased, the stocks rose faster, and 25-year-old billionaires became an everyday news story. And this is how it all unfolded:

- Venture capitalists had been putting too much money into dot-coms. Since these dot-coms needed a place to host their web sites, Web-hosting companies like Exodus Communications were formed and consumed gobs of bandwidth.
- At the same time, companies that promised to deliver high-speed Internet access to consumers using digital subscriber line technology (like Covad, NorthPoint Communications, and Rhythms Net-Connections) were beginning to grow quickly. These companies needed bandwidth to provide to their growing customer base.
- Thus Level 3, Qwest, and Global Crossing—all providers of backbone networks to corporations and other phone companies—were born. These fast-growing networks required equipment.
- Enter the telecom hardware makers, like Lucent, Nortel, Cisco, and eventually a host of start-ups that raised billions of dollars from venture capitalists eager to unseat the equipment-making incumbents.

Most of these companies were starting from a very small base, so it created a perception that the demand for bandwidth was growing and was endless. Thus was born the Internet myth that "Internet traffic doubles every 100 days," an urban legend that resulted in the mad rush of money into new start-ups, whether makers of telecom equipment, cable modems, or broadband service providers. But the notion of Internet traffic doubling every 100 days was hogwash—by 1997, Internet traffic actually only doubled once each year.

Shamans, Analysts, and Other Boom-Time Heroes

The broadband bubble and the dot-com bubble resulted from overblown expectations and irrational exuberance. Investors could relate to dot-coms, whose technology was transparent; it's pretty easy to comprehend buying books over the Web. The broadband sector's technology boasts, by comparison, were opaque. Average investors and even many savvy investors didn't understand what made broadband possible. So they relied on pundits, analysts, and the media to translate the technological mumbo jumbo into plain English. Enter analysts like Jack Grubman, pundits like George Gilder, and market research firms with their ever-bullish forecasts.

Even while the dot-com sector began to hemorrhage in March 2000, the broadband bubble just kept growing. You would have thought investors would have learned something from the dot-com bubble, but they didn't. They took their money out of dot-coms and, this time, fueled the broadband bubble.

No one saw fit to fly cautionary flags, and in this respect I, too, am to blame. I think the business media, of which I am a member, never entirely caught on to the house of cards that Jack Grubman and his broadbandit gang were building. For the longest time, it seemed there wasn't a thing that the broadband companies could do wrong. This all-or-nothing hype and the cavalier recommendations succeeded in creating a frenzy that has been unparalleled in the history of technology and the capital markets.

Grubman was touting the companies in exchange for banking business, and CEOs at many companies were busy taking IPO shares in exchange for this business and getting richer. In the 1990s, the IPO market was like heroin. Grubman was the dealer, and the corporate titans the addicts. But Grubman was merely a symbol of the crumbling ethical value system. Morals were sacrificed in an attempt to make easy money. The losers were the small investors, who hadn't a hope in this rigged game.

Conflicts of interest were part of the greed culture that permeated Wall Street and corporate America. But no one paid any attention to this, for America was in the grip of a bull market like never before. Unknown dot-coms were raising billions of dollars from the market. Folks like John Sidgmore of UUNet were talking up the demand for Internet bandwidth, and it seemed a new communications revolution was unfolding in front of our eyes.

New technologies, the World Wide Web, the Internet, e-mail, and fiber optics made everyone myopic. It seemed that stocks could only go up and markets defied gravity. It was a world where Jack Grubman and George Gilder were as close to deities as mere mortals could get.

It's the (New) New Economy, Stupid!

Build it (the networks) and they (the customers) will come, preached King Jack Grubman to his many disciples in the telecom world, who took him seriously. What they didn't know was that Jack Grubman was not all that he was cracked up to be! Despite his brief stint as an analyst for AT&T, Jack Grubman knew nothing about sales, how a business was run, or even the basic principles of customer behavior. Grubman got his ideas from those who ran the businesses, but they in turn took advice from him—a mutually parasitic relationship. In the end, nobody had a clue.

And just when there was no real reason to push the stock, Wall Street would come out with yet another market-driven, dead-wrong metric. Take Winstar and Teligent as examples: Analysts would count their building leases as a part of growth. How did that work? Access to the buildings where Winstar and Teligent could sell their services meant more tenants, that's how. So Teligent signed leases and built a network, but it could never sign up enough customers to have meaningful revenue. Many of these companies created an illusion of revenue growth, using a technique that later came to be known as multiple counting. This helped them raise money from the markets and grow bigger—except they could not really find enough customers.

Dude, Where's My Bubble?

Since writing this book, I have been asked this question: Why did this happen? After months of contemplation and talking to several very smart people, I have come up with a few reasons.

1. **Fear of the falling stock.** Most chief executives were scared to stand in front of the Wall Street community that was valuing them on revenue growth and admit that the growth was not there. It would have meant taking a big hit on stock price. Yet such moves could have prevented the bankruptcies of WorldCom and Global Crossing. No one seemed to take responsibility for the companies they worked for; all they were responsible for was their own skin.

2. **The cult of the CEO.** In the 1990s we made the CEO the new American hero. We put him on the cover of magazines, we celebrated his lifestyle, and we treated him like Jerry Rice and Alex Rodriguez. CEOs forgot that they served the shareholders and not their own bank accounts.

3. **CEO salaries and stock options.** According to *BusinessWeek*, the average CEO made 42 times the average hourly worker's pay in 1980, 85 times it in 1990, and 531 times it in 2000. And that did not include the liberal stock option plans that had very little downside for chief executives. Why would CEOs worry about anything other than boosting the stock price—because the higher the stock prices, the richer they got. No wonder they did not admit to Wall Street that business was not so good.

4. **Hypergrowth.** Our society came to expect 20 percent annual returns on investments in an economy that typically grows only 3 percent.

The pages that follow tell the stories of a handful of men, a few dozen companies, and a culture that worshiped at the altar of greed. This is the story of a plague of myopic avarice that struck in a manner not unlike the blindness that decimates a country in Jose Saramago's Pulitzer Prize–winning novel *Blindness*. This is the story of insane optimism, astonishing selfishness, and a series of colossal failures to obey the tenets of

Economics 101—all leading to the empty 401(k) accounts of people like Paula Smith. These are the stories of the Broadbandits.

The stories have been collected from various sources including published reports, which I have tried to list as accurately as possible. In order to write these stories I interviewed more than five dozen industry insiders, in addition to using material from my reporters' notebooks. However, all impressions are my own.

Om Malik
December 1, 2002

PART I

THE FIBER BARONS

1

BERNIE'S
BAD IDEA

Bernie Ebbers, the founder of WorldCom, had made a very bad bet and lost. The broadband business he had so ruthlessly built with other people's money was being billed as the biggest corporate con in the history of America—even bigger than Enron. Ebbers had made the fatal mistake of believing his own hype, even as he bilked investors of their money.

It was March 2002 when it all started to unravel. The Securities and Exchange Commission (SEC) had finally caught on to the shenanigans at WorldCom and started an investigation. In a month, Ebbers would resign. Ebbers' coconspirators, WorldCom Chief Financial Officer Scott D. Sullivan and Controller David Myers, had been grilled by the Internal Audit Committee and would ultimately be fired from the company. On August 1, they would surrender to the police, knowing they faced at least five years in prison, hefty fines of $250,000 and several counts including securities fraud, conspiracy to commit securities fraud, and filing false statements with the SEC.

Bernie Ebbers, king of bandwidth, telecom titan, and corporate conqueror like none before, could be next. Every step that Sullivan and Myers would take to the police car waiting for them outside New York's posh Hotel Elysee would take Ebbers farther away from his empire.

"Fifteen unbelievable years and two very challenging years" is how Bernie Ebbers would describe his tenure at WorldCom on his 17th anniversary at the company on April 30, 2002.[1] That would also be his last day at the company he had started in a Mississippi coffee shop in 1983.

■ ■ ■

Ebbers' story is classic American fare: an immigrant from a tiny town who made it big but in the end became the victim of his own success. Had he been manipulated by Wall Street shysters, or had the money been corrupting his soul one step at a time?

Ebbers' life began 60 years ago in a working-class neighborhood in Edmonton, Alberta, in western Canada. He attended Victoria Composite High School in Edmonton and supported himself through high school by working as a milkman and as a bouncer at a bar. It seemed like he was going to have an average working-class life, when his luck turned. Mississippi College, a small Baptist college in Jackson, Mississippi, offered the six-foot-four Canadian a basketball scholarship. He accepted, went to Jackson, got an undergraduate degree in physical education, and settled down in nearby Brookhaven. He worked as a basketball coach first, and then invested in a motel in Colonel, Mississippi, in 1974, which, after some hard work and savvy deal making, grew into a collection of Best Westerns in Mississippi and Texas. Along the way, he also acquired a car dealership in Columbia, Mississippi. By 1983, nine motels in the two states were controlled by his company, the Master Corporation.

At that time AT&T, then the largest phone company on the planet, was in the process of breaking up into AT&T and seven local phone companies. The breakup was the result of a federal action prompted by lobbying and efforts of MCI, then a small long distance provider. Now anyone, however small, was free to compete for the phone dollars. Recognizing the winds of change, in 1983, Ebbers got together with four investors to fund a long distance reseller. WorldCom folklore has it that Ebbers and the three other investors, Bill Fields, David Singleton, and Murray Waldron, hatched the blueprint of the company at a Hattiesburg coffee shop. The company's original name was apparently suggested by a waitress who inquired about what kind of company they were planning. According to WorldCom's 1998 annual report, "[the] waitress walked away for a few minutes and came back with a napkin on which she had scribbled the letters "LDDC"—Long Distance Discount Calling. Later the name was changed to Long Distance Discount Service (LDDS).

"We got in as passive investors, saw the gold at the end of the rainbow, and thought it was going to happen the next day. It didn't work out that way," Ebbers later told the *Jackson Journal of Business*.[2] Instead, the company lost large sums of money in the first two years. The investor group decided to put Ebbers in charge. He was the perfect choice: A penny-pincher, he got the company to focus on small business customers who spent less than $20,000 a year on long-distance phone calls. The big guns like AT&T and even MCI ignored this segment of the market, focusing instead on Fortune 500–type companies that ran up huge phone bills. Ebbers figured it was easier to nibble at the long distance market in untapped segments than to take on AT&T and MCI from the word go. By employing this strategy, his company was able to stay just below the radar and survive on telecom crumbs for the next decade. But soon mere crumbs would not be enough to feed Ebbers' increasing appetite.

After the AT&T breakup, several long distance discounters had the same idea as LDDS; they, too, had set out their shingles and gotten into the telecom business. But, like LDDS, most of them were struggling because they either lacked hardware expertise or were run by telecom engineers with no marketing and sales savvy.

Ebbers quickly figured out that his company, which had both of these shortcomings, needed to get outside help. He found engineers and outsourced the hardware side of the business, but he still needed a marketing whiz. At a trade conference, Ebbers happened to met Diana Day, a former copywriter for a local television station who had once also done public relations for the Mississippi Republican Party. Day was working for a telecom reseller and was bored out of her mind when she met Ebbers, who decided to give her a shot at marketing. Turns out it wasn't a bad bet. Day turned things around for WorldCom: Sales skyrocketed from less than $1.5 million in 1984 to $3.3 million in 1985 and $8.6 million in 1986.[3]

Around this time, Ebbers came up with another idea—growth through acquisition. It was a fairly simple strategy: Buy a company, keep its sales team, absorb its customers, and use LDDS's cheap infrastructure. By 1987, Ebbers' LDDS had bought five companies, mostly in the South. It was a perfect way to boost sales and become a big fish in, admittedly, a small pond, while staying away from AT&T, MCI, and Sprint.

Bernie focused on customer service. He knew that if LDDS paid attention to its customers, it wouldn't be long before more business would flow his way. While large companies typically ignored the small customers and treated them with scorn, customer service was his company's hallmark in the beginning, with the company providing affordable long distance to everyone. It soon seemed that Bernie Ebbers was on his way to becoming the Sam Walton of the telecommunications industry.

In 1989 LDDS acquired the publicly traded long distance reseller Advantage Companies of Atlanta, Georgia, and through the acquisition, LDDS became a public company. Also as a result of this acquisition, a key partner in future shenanigans, Stiles A. Kellet Jr., came into Ebbers' supernova. Kellet was the chairman of the board of Advantage and became an LDDS board member after the merger. In later years at World-Com he would play a vital role in the great worldcon.

Ebbers was now in contact with Wall Street and its financial analysts. One of them was Jack Grubman, then an analyst at the New York investment bank Paine Webber, who was eventually to play a major part in the rise and fall of Bernie Ebbers. "We both come from the wrong side of the tracks vis-a-vis the financial community. And I can relate to him far better than most people I deal with. Bernie and I would have a strategic session in Jackson, and it usually was while shooting pool and drinking beer," Grubman would later say.[4]

Then, in December 1992, LDDS acquired another long distance reseller, Texas-based Advanced Telecommunications, for about $850 million. Ebbers' life was about to get complicated. This acquisition brought with it a man called Scott Sullivan. When LDDS bought Advanced, Sullivan signed on as assistant treasurer for LDDS and was promoted to chief financial officer of the company by 1994. At about the same time, Jack Grubman quit Paine Webber and joined Salomon Brothers.

If Grubman became the one who encouraged Bernie to grow bigger through acquisitions, then Sullivan was the one who massaged the balance sheets to make it happen. The unholy troika was complete. Ebbers soon went on an acquisition tear, and the "Bernie and Scott show," as Wall Street dubbed it, had begun. Bernie would become the face of the company, Scott would bleed the acquisitions dry, and Jack would be the shill who would tout the company's stock. LDDS shares started to go through the roof, giving Ebbers more currency to continue his shopping spree.

To the outside world, Ebbers was becoming the monster with the insatiable appetite for telecoms, but in reality it was Sullivan who did the grunt work and was the real force behind all those deals. While Ebbers pranced around like a cowboy, Sullivan was busy explaining the various mergers to Wall Street. A former General Electric executive and Klynveld PeatMarwick Goerdeler (KPMG) accountant, he was quiet, had a way with numbers, and was the exact opposite of Ebbers' aggressive, in-your-face personality. Had it not been for Sullivan, Ebbers would not have had the gumption, or the knowledge, to sell the complicated mergers to Wall Street.

Surprisingly, the two men got along famously despite their contrasting personalities. They had adjoining offices and were often seen having lunch at the company cafeteria. Ebbers was a very suspicious man and a bit of a control freak. He never hired a chief operating officer and instead relied on a mere accountant to help run his business—perhaps because he didn't want anyone to know the real facts. No one dared question Bernie and his ways. But he trusted Sullivan.

In December 1994, Ebbers bought IDB WorldCom, an independent long distance phone company, for $936 million, and changed his company's name to WorldCom—a more grandiose moniker befitting its new stature. A year later WorldCom bought Williams Technology Group, also known as WilTel, a network carrier, for $2.5 billion. By the end of 1995, WorldCom had sales of $3.9 billion and debt of $3.4 billion—the debt-to-revenue ratio was perilously high, and would remain so for the entire history of the company. But even then, no one on Wall Street dared question Bernie. He would stride into meetings with investors, his chief financial officer Sullivan at his side, put up a graph showing WorldCom's share price headed up, and smugly ask: "Any questions?" Ebbers got more audacious with every acquisition. For a while, it seemed there was no price he couldn't afford.

St. Bernie Rises

Ebbers' rise to the top and his newfound riches made him the new savior of Mississippi, one of the most underprivileged states in the country. He raised funds for local colleges, donated to charities, and employed thousands. He taught Sunday school at his Baptist church, served meals to the homeless, and helped get an economic revival going in Mississippi. In his adopted hometown, he became St. Bernie!

The company built a shiny new office in Clinton, a town of 24,000 quite close to Jackson, the state capital. Ebbers' company hired interns from local colleges and became a major sponsor of civic activities such as arts and crafts fairs and street festivals. He may have been a billionaire, but to the people of Mississippi he was one of them. Once he vowed that he would leave the state "only in a box." He drove an old, dented Ford truck that had a loose fender, wore blue denims and boots to work, and even wore a pair of jeans with a tuxedo jacket to his second wedding in 1999.

People would often see him walking about town chomping on a cigar. He would play pool, drink beer, and chat with locals as Willie Nelson music played in the background. Ebbers became so popular in the Jackson-Clinton area that people mentioned him as a possible gubernatorial candidate. But his long, flowing hair and homilies made him seem more like a country-and-western singer than a top chief executive or a possible governor. Ebbers' act worked especially well with investors. They quickly developed a cultlike following for their hero, who would say things like "I'm no technology dude."

But behind the folksy façade, Bernie cultivated a taste for the good life. He was so caught up in his image as a takeover tycoon that he bought a $60 million yacht and named it *Aquasition*. He got other trappings of success as well, such as a stake in a minor league ice-hockey team, the Bandits, and a 900-acre farm that included a 100-acre lake. He built a million dollar mansion and a stone-cedar lodge. In 1998, he acquired a 164,000-acre ranch north of Vancouver, British Columbia, for $68 million. The ranch had 11 lakes, 22,000 head of cattle, some 300 horses, and sundry businesses.[5] It was almost the size of the state of Rhode Island. According to published reports, Ebbers not only bought the ranch, but he also bought timberland in the American South and a yacht-building company in Savannah, Georgia.

Along with the wealth came arrogance. He threw out a local reporter at a stockholders' meeting because he didn't like what the scribe had written.[6] While he publicly preached Christian values, it is said he would stay up late drinking with executives at WorldCom. He left Linda, his wife of three decades, and openly courted a WorldCom saleswoman, Kristie, who was also married at the time. "It's so inappropriate the way he wears his Christian banner. The people down here refuse to accept the hypocrisy," said a local Jackson real estate agent of Bernie's antics.[7]

Increasingly, it seemed like Bernie's once folksy aspirations were taking a sharply more ambitious turn. With the advent of the technology boom in the mid-1990s, Bernie saw his chance to trade up from mere telecom crumbs to the meat and potatoes of the next telecom revolution: broadband. With all the hype surrounding the Internet's potential, the fact was that most people were stuck in a world of dial-up modems. The vision was that this narrow, slow channel would soon be replaced by a much bigger pipeline over which one could get blazing fast access to CNN.com, listen to music, or even watch movies. This high-speed connection, 50 times the speed of a slow modem, is called a broadband connection. Broadband connections need big infrastructure, the kind Ebbers was slowly acquiring.

In fact, WorldCom had stealthily become the quintessential example of an old-style phone company that got a broadband makeover. Gone was the company's customer-centric focus of yore. In a full-scale about-face, Ebbers baldly admitted that his customers were now his investors. He told *BusinessWeek*, "Our goal is not to capture market share. . . . Our goal is to be the No. 1 stock on Wall Street."[8] WorldCom shareholders loved Ebbers, and he became the king of takeovers. By the time Ebbers got the boot in 2002, he had bought about 70 companies and pushed the stock up 7,000 percent, (before it became worthless paper), making many people very rich, including himself.

Masters of the Internet

One of the driving forces in WorldCom's rising stock price was UUNet, a pioneering Internet provider that the company acquired in 1996 through its acquisition of MFS Communications.

UUNet was the brainchild of Richard Adams, a Cleveland native and an alumnus of Purdue University. Adams used to work for a Defense Department contractor called the Center for Seismic Studies (CSS), in Arlington, Virginia. CSS was connected to Arpanet, a government-owned network and a precursor to the modern Internet. Adams also became part of an organization called USENIX Association that was formed by computer geeks in Virginia and nearby Washington, D.C. What united the group was a love for UNIX, a type of high-end software operating system and programming language. Adams told the *Washington Post* in 1996 that most of them wanted to get connected to Arpanet mostly to exchange e-mails. "I do remember thinking, if people were begging you to take their money for this, that's the business I want to be in," Adams said.[9]

The association pooled together $250,000 and asked Adams to put together an experimental network that could provide USENIX members with some links. In 1987 this became a not-for-profit company that Adams named UUNet, after the UUCP, the Unix-to-Unix Copy Protocol. Eventually, in 1989, Adams quit his job, took the company out of USENIX, and moved it into his home. He had four employees and about $1 million in revenue, mostly from USENIX members. A year later UUNet became a for-profit company.

In 1992, a chance meeting with Mitch Kapor, the founder of the software company Lotus Development, resulted in the first investment in the company. By then UUNet was making about $500,000 in profit every year on sales of over $6.4 million. The pesky little start-up was getting attention. MFS Communications, a small local phone company started by James (Jim) Crowe (who would later start Level 3 Communications), came calling.

Jim, the son of a decorated World War II hero, had worked for the Omaha-based construction giant Peter Kiewit Sons. There he met Walter Scott, the company's chief executive, and in 1988 convinced Scott to back MFS. Scott put about $500 million into the company, which went public in 1993. MFS had been wooing UUNet for a while and had made an $8 million buyout offer, but Adams demurred and held out for a better price. In five years, Adams' stake would be worth $560 million (according to *Forbes*), though he didn't know that at the time.

Meanwhile, Kapor's investment resulted in the influx of more venture capital money—about $12 million from Menlo Ventures, Accel Partners, New Enterprise Associates, and Hancock Venture Partners. Adams' share fell from 98 percent to 15 percent of the company. The new investors insisted on more professional management, but that resulted in a mutiny. It was a classic suits-versus-geeks battle. The geeks, led by Michael O'Dell, who had joined UUNet earlier as the chief technology officer, were experts in networking technologies. They were the only ones who knew how to keep UUNet running, and they ultimately won the battle.

Hoping to calm things down, UUNet's backers hired John Sidgmore, a General Electric veteran, as chief executive officer. Sidgmore, who had spent 14 years at GE Information Services, had recently sold Intellicom Solutions, a telecom software company, for $100 million to Computer Sciences Corporation. He accepted UUNet's offer and joined as CEO in 1994.

The company shifted its focus from retail customers to corporations and began building a worldwide network that could service large corporations like Wall Street banks. Since most of these corporations had global operations, they needed the means to exchange data and communicate with each other. UUNet would install special phone lines that would handle data traffic at lightning speed.

UUNet also became the company that would provide the backbone for Microsoft's much-vaunted online service, the Microsoft Network (MSN). MCI, AT&T, and other smaller rivals that had started copying UUNet's business strategy were left agog. As part of the MSN deal, Microsoft bought 17 percent of the company. UUNet and John Sidgmore had arrived.

There was no one who could speak with more authority about the Internet than Sidgmore and his geeky sidekick Michael O'Dell. After all, they were the first ones to arrive at the commercial Internet, and they were building the biggest and the most cutting-edge network in the world! Everyone acknowledged their leadership.

Three months before Netscape's historic IPO, UUNet announced its initial public offering, underwritten by Goldman, Sachs & Co. On May 25, 1995, the stock was offered at $14 a share, and closed the day at $26 a share, having reached $28 a share in intra-day trading. The company raised $68 million, and John Sidgmore was richer by about $34 million. The subsequent Netscape IPO in August 1995 and ever-bullish analysts pumped the stock market higher.

The fascination with the Internet was just getting started. UUNet went on an acquisition binge in Europe, hired hundreds of people, and the stock hit $93.25 a share in November 1995. Soon UUNet was gracing the covers of magazines, and its employees became the icons of the new gilded age. In October 1996, *Forbes ASAP*, the now-defunct technology supplement of

Forbes magazine, featured a photo spread on the company, 40 of whose 700 employees were millionaires.

Ahmar Abbas remembers the halcyon days well. Now a principal at the research firm Grid Partners, he joined UUNet in May 1996 in the company's sales department. He had previously worked at Salomon Brothers, which was an early UUNet customer. There were 130 employees who worked at UUNet then, and "it was so small that you did everything," he recalls. The big moment came when UUNet signed a deal with Microsoft to build out its dial network.

Abbas called the pace frenetic. When he joined, the company was selling ten high-speed connections, known as T-1s, per month, and when he left four years later the number had risen to 2,000. The number of dial-up ports—points that most of us connect to when we try to log on to the Internet using, say, America Online—increased from 100,000 to 4 million, and the company expanded to 28 countries. But no party lasts forever, and in 1996 new competition emerged in the form of AT&T and MCI, both of which started to chip away at UUNet's almost-dominant market share by launching their own Internet initiatives. Some of the wind went out of UUNet's sails.

In addition, other competitors such as PSINet of Herndon, Virginia, and Netcom On-Line Communications of San Jose, California, were getting a lot of play in the media for their consumer-centric approaches, and their stocks were on an upswing as a result. People had assumed UUNet's dominance of the Internet access business, but tough competition made Wall Street ask tough questions. And when AT&T announced its WorldNet Internet access service, investors soured on UUNet. The company had to shelve its secondary offering, and by March 1996 the stock had sunk to the mid-$20s per share. Sidgmore needed money to complete the UUNet network and to expand, if he was to stay ahead of the game. "It was clear that we needed to go with a big company, for we didn't have the money to build out the infrastructure," recalls Abbas.

Sidgmore knew the game was up. Soon after joining UUNet, Sidgmore had tried unsuccessfully to merge the company with Netcom. Now, when MFS Communications showed up again, this time with a $2 billion buyout offer—a 100 percent premium over the UUNet stock price at the time—he decided to accept it. After all, MFS Communications had one of the best-performing stocks in the market. On May 1,

1996, UUNet was sold. Sidgmore and O'Dell still had solid reputations. They had become the gurus of the broadband world, and their words—"Internet traffic is doubling every 100 days"—became the gospel of the new economy. Even the U.S. government quoted their line in its reports.

The Great Internet Myth

Think of "Internet traffic doubles every 100 days" as an urban legend—along the lines of the Neiman Marcus cookie recipe. One of the many things the Internet has helped Americans exchange is urban legends, and a classic example of these myths, which are usually sent via e-mail, is one that tells the supposed story of a Neiman Marcus customer who accidentally bought the company's chocolate chip cookie recipe for $150. The customer allegedly took revenge on the company by distributing the recipe via e-mail.

Just like that myth that has circulated for years, during the broadband boom of the 1990s, "Internet traffic doubles every 100 days" was a legend that, instead of bringing ordinary citizens to write angry letters to a department store chain, convinced executives, analysts, venture capitalists, and retail investors to pour billions of dollars into telecom companies. The belief in this legend resulted in the mad rush of money into new start-ups, be they makers of telecom equipment, cable modems, broadband service providers, or even book retailers such as Amazon.com. Even otherwise conservative venture capitalists poured billions into broadband companies. Wall Street also got into the act and sold bad business ideas to the unsuspecting masses.

But Andrew Odlyzko, a research scientist for AT&T Labs in Florham Park, New Jersey, didn't buy it. In 1997, Odlyzko and his colleague, Kerry Coffman, decided to undertake an academic exercise to analyze data available from AT&T and other major Internet providers such as MCI and BBN Planet. UUNet, which by then had been acquired by WorldCom, kept its information under lock and key. Odlyzko and Coffman spent almost a year analyzing the traffic patterns on the Internet and wrote a paper, "The Size and Growth of the Internet," which was released to the public in October 1998. Their findings proved that the whole notion of Internet traffic doubling every 100 days was hogwash. The duo had found that Internet traffic was only about doubling each year, or, more precisely, that it was growing at between 70 and 150 percent a year.

No one wanted to believe them, and they were dismissed as two musty academics working for AT&T, a company that was scandalously late to the Internet business. Even their corporate employer didn't take them seriously. "I was discounted by everyone and the AT&T management was not supporting me, so it was no surprise that the WorldCom hype machine succeeded," said Odlyzko, who quit AT&T in 2001 and now works as the director of the Digital Technology Center at the University of Minnesota in Minneapolis.

"Over the years, every time I tried to trace the rumors of 'Internet traffic is doubling every three or four months' to their source, I was always pointed at folks from WorldCom, typically [Bernie] Ebbers or [John] Sidgmore. They, more than anyone else, seemed to be responsible for inflating the Internet bubble [until it collapsed on them]," Odlyzko noted in an e-mail.[10] "Many times, inside AT&T, when I would bring up the studies Kerry Coffman and I had done showing that Internet traffic was only about doubling each year, this work would be dismissed with remarks of the type, 'But Mike Armstrong was at XXX last month and heard Bernie Ebbers say explicitly that UUNet traffic is doubling every 100 days. We just have to try harder to match those growth rates and catch up with WorldCom,'" Odlyzko wrote in that e-mail message.[11]

Every time Odlyzko put up an argument, it was like screaming into a 100-mph gale-force wind. And even if he did get a chance to get a word in, it would immediately be countered by someone from WorldCom and its UUNet division. Odlyzko and Coffman had no chance of winning the argument. They were speaking in bits and bytes while the others were spouting the new lingo of billions, which was much more palatable.

Thanks to some great media positioning and public relations strategy, UUNet had been deemed the "Internet authority." UUNet executive John Sidgmore and his trusted lieutenant and chief scientist Michael O'Dell were the St. Paul and St. Peter of the Internet religion. Whatever they said must be true, everyone reasoned. Their sayings were just the kind of mantra the broadbandits needed to raise capital and justify their spendthrift ways. WorldCom's hype machine would issue press releases with statements like this one: "With dial access demand growing at the rate of over 10 percent every week and traffic over the backbone almost doubling every quarter, UUNet Technologies, Inc., the largest provider of Internet services in the world and a subsidiary of WorldCom, Inc., announced today that it has initiated the most ambitious network expansion plan in the company's history."[12]

In an interview with *Telecommunications* magazine in August 1997, Sidgmore touted the hypergrowth model. "We're seeing growth at an unprecedented level. Our backbone doubles every 3.7 months, which means that it's growing by a factor of 10 every year. So three years from now, we expect our network to be 1,000 times the size it is today. There's never been a technology model with such an extraordinary rate of growth like this before. So the big challenge is to deploy infrastructure fast enough to accommodate such a growth rate. We're in a supply-constrained economy for the first time in the telecom industry," he said.[13]

"Sidgmore came up with the Internet number and everyone took him at his word. He knew that UUNet bandwidth was not growing at the speed that they were telling the customers and everyone else," says Roy Bynum, who came to WorldCom in 1998 as part of the MCI acquisition. Bynum, one of the earliest champions of the Internet at MCI, along with a few dozen employees, had made MCI a worthy competitor to UUNet. "Yes, there was a period of time when commercial Internet traffic doubled every 100 days, but that is true of every new network service [since it starts from a near-zero user base] such as ATM or Frame Relay, when it was first created," Bynum continues. "Internet traffic was doubling every 100 days only during the initial commercialization and the initial rollout of the new HTML Web browser, Netscape, in 1995 and 1996 and, to some extent, in 1997.

"After 1997 and the initial surge, the whole spiel about Internet traffic doubling every 100 days became a scam. It was a very positive scam, so there were a lot of people who wanted to believe it because they could potentially make money from it," added Bynum. "Ebbers and Sidgmore were presenting to everyone that bandwidth at the core of the network was growing, but that was simply not true."

Despite all the talk about broadband and the need for fatter pipes, it was the slow and unfashionable business of dial-up Internet access that was supporting the revenue base at UUNet. It is rumored to have accounted for nearly 50 percent of total revenues. "The whole Internet protocol side of the business never really made money, and it was the dial-up network that was making money," recalled Abbas.

That might explain why MFS's head honcho, Jim Crowe, decided to sell the company to WorldCom a scant three months after it announced it was acquiring UUNet. For the first six months of 1996, MFS had a net loss of $193.5 million on revenue of $416 million. They needed to flip the company—fast. So when Bernie Ebbers came to town and offered to buy it, Crowe didn't refuse his $14 billion offer. Ebbers was looking for Internet pizzazz. The long distance business, thanks to the brutal price wars, wasn't the cash cow it had once been, and WorldCom desperately needed to boost revenue to keep its share price high. Higher share prices would give Ebbers the currency to continue his expansion binge. He had to keep gobbling up rival companies or his house of cards would come crashing down. So, after MFS, his next target became MCI.

Battle for MCI

MCI started in 1962 in Springfield, Illinois, under the name Communications Consultants. Its goal then was to sell General Electric's radical new two-way radios. The partners in this enterprise were Jack Goeken, a General Electric representative and owner of Mainline Electronics; Donald and Nicholas Philips; Leonard Barrett; and Kenneth Garthe. Since Springfield didn't have a GE dealership, the quintet decided to make the Illinois capital the headquarters.

Initially, the business did better than they expected, but they ran into trouble when another GE representative complained that they were selling radios in his territory, and GE withdrew Communications Consultants' license. Goeken then came up with the idea that the company should build a microwave network—that is, a string of microwave transmitters and receivers—to connect St. Louis to Chicago. The route between the two cities was a heavy trucking route, and a network would help the truckers continuously use their radios to stay in touch with people at their home base. On October 3, 1963, they started Microwave Communications Inc., or MCI. But in order to build the network and offer the service, Goeken needed permission from the Federal Communications Commission.

The service MCI planned to offer looked like phone service to AT&T. MCI's microwave technology could easily replicate the AT&T network, and that worried Ma Bell. It was a threat to its monopoly, and AT&T did everything it could to crush the upstart. By 1964, MCI's finances were in the doldrums and Donald Philips decided to bow out of the partnership. The company would be dragged through legal battles for the next few years, each year pushing it closer to the financial brink. They had no FCC approval, no money, and no network.

In 1968, Bill McGowan, a venture capitalist of sorts, heard about MCI. McGowan had grown up in a coal mining town in Pennsylvania, gone to college on the GI Bill, and later attended Harvard Business School. He loved the idea of taking on AT&T. He had connections on Wall Street that would help with the funds to build a network when and if MCI got FCC permission.

McGowan offered $35,000 for 50 shares of MCI. He convinced MCI's shareholders to incorporate a new company called Microwave Communications of America, Inc., in Delaware. The ownership was split four ways between McGowan, Goeken, MCI, and a new entity that eventually became Micom. In 1974, Goeken stepped down as president (he left MCI three years later), and McGowan was put in charge of operations. A new era was beginning, but it would still be years before MCI could raise money, build its network, and offer its services to customers.

McGowan desperately wanted to unseat AT&T's monopoly and make MCI a profitable company. The story goes that he kept a cracked AT&T cement block logo in his office, just to remind him of his real objective in life. His efforts eventually did result in AT&T being declared a monopoly and in its subsequent breakup in 1984. By then, MCI had employed so many attorneys to duke it out with Ma Bell that telecom wags described the company as a "law firm with antennae."

The company, now free to compete, built a nationwide network, thanks in part to liberal funding raised for the company by junk bond king Michael Milken in the 1980s. An aggressive sales strategy combined with a weakened AT&T was all McGowan needed to turn MCI into a multibillion dollar company. Its long distance phone service offerings proved a hit with corporate customers. MCI built a nationwide fiber-optic network that was more efficient than Ma Bell's networks, and this helped MCI offer even better prices to its customers. Sales boomed. McGowan was a great motivator and innovator, and he nudged the company toward making new offerings, including data services.

But with McGowan's tragic and sudden death in 1992, MCI, which used to thrive on its gonzo culture and devil-may-care attitude, started a downward slide. MCI never really recovered from McGowan's death. He had been the motivator for the entire company. Jerry Taylor, MCI's president, and Bert Roberts, MCI's chairman and chief executive, were McGowan's lieutenants, but the MCI army needed a General Patton. The long distance wars with AT&T flared up again, largely because of the efforts of Joe Nacchio, a young, cocky AT&T executive who matched MCI move-to-move.

Nacchio, who later left to become the chief executive of Qwest Communications, knew that AT&T's cash position was much stronger than MCI's, as the latter was getting crushed under the weight of the interest payments on its debt. He waged an economic war with MCI and won. MCI had also missed out on the wireless boom and didn't fully capitalize on its Internet service provider business. "There was no edge to the company, the energy was down, and it was clear to us, MCI was going nowhere," recalls Phil Jacobson, a former MCI executive and now a principal with the networking consultancy Network Conceptions. For almost two years, MCI tried to become a leader in the Internet business but was out-hyped by UUNet.

In 1995, MCI started getting involved in money-losing ventures like a toll-free service for people to listen to music before buying it. The short-lived 1-800-MUSIC-NOW service, which was shut down in 1996, cost MCI about $40 million. The company also made a disastrous foray into the Internet content business and lost close to $50 million on that project. "Everything was beginning to stagnate, and it was quite different from early days when the company was focused on taking down AT&T," recalls Jacobson. "Managers were just hanging around, the stock was sinking, and the company was just becoming very political. Backstabbing was routine," recalls Jacobson's partner Farooq Hussain, who worked in the Internet group for MCI. "McGowan did not groom anyone to take over."

MCI executives wanted to sell out, and in November 1996, British Telecom (BT) offered $24 billion for MCI. But even before the final agreements could be drawn up, there was dissension in the ranks. The executives weren't happy. British Telecom apparently didn't understand the complexities of the merger. "It was all about idiotic concerns, about empire building and management control, and not about business expansion," recalls Hussain, who worked on trying to combine the two companies.

The British Telecom bid and the ensuing negotiations sent a clear signal that MCI was ready to throw in the towel and would sell itself to anyone with a decent offer. Jack Grubman told his clients that BT was getting a bargain. On the other side of the pond, BT executives were being laughed at for paying too much. The *Financial Times* reminded them of their stupidity on an almost daily basis and wanted the British monopoly to call off the deal. BT lowered the offering price from $24 billion to $19 billion. This led to another round of negotiations. Then, extraordinarily, in October 1997, WorldCom made a $30 billion hostile bid for MCI. The business world was shocked. MCI was almost four times the size of WorldCom, which at the time had sales of merely $4.5 billion, while MCI's sales were $18.5 billion. A minnow wanted to swallow the whale!

For WorldCom executives it was the deal of a lifetime. If they could pull this one off, WorldCom could be among the top ten phone companies in the world, with revenues that could help pay off nearly $5.4 billion in long-term debt. "I thought about BT's position for making a lower bid and the marketing and cost advantages that favored World-Com to make a successful higher bid. I was convinced that we could unseat BT as the incumbent acquiring company to a position where they would not be able to bid against us," boasted Sullivan in an interview with *Oswego*, a magazine published by the alumni association of Oswego State University, Sullivan's alma mater. Meanwhile, WorldCom decided to make a $2.5 billion bid for Brooks Fiber, another network operator.

Unfortunately, another local phone company, GTE, also decided to make a bid for MCI, and WorldCom had to increase its offer. GTE's bid started a feeding frenzy and the telecom industry, normally staid and boring, was once again in a state of upheaval. It was like the junk-bond–driven 1980s. Bernie Ebbers was the new hero on Wall Street. After all, his machinations were making investment bankers very rich. Lazard Freres & Company was counseling MCI. N. M. Rothschild of London and Morgan Stanley, Dean Witter, Discover & Company were in the British Telecom camp. Jack Grubman and Salomon Brothers were on WorldCom's team. Goldman, Sachs & Company and the Bear Stearns Companies were advisers to GTE. Lehman Brothers later joined the MCI team.

"Telecom in play; as British Telecom, WorldCom and GTE vie for MCI, merger frenzy has gripped this once staid industry," gushed *Fortune*.[14] Ebbers remained arrogant and condescending toward MCI throughout the fight, which lasted almost six months. In the end he won, but he paid dearly. It cost him $37 billion to buy MCI. The combined companies had $28 billion in annual sales. In order to finance the purchase, WorldCom took on about $20 billion in debt, got a $12 billion bank facility from NationsBank and another $3 billion in a commercial paper program through Lehman Brothers. As part of the deal, about $7 billion of the total went to BT shareholders—in cash. Having snapped up Brooks Fiber, UUNet, and MFS, Bernie had one little thing left to do: buy the network businesses of CompuServe and America Online so he could finally become a bandwidth baron.

In order for the U.S. Department of Justice to approve the MCI/WorldCom merger, MCI was required to sell off its Internet division, internetMCI. Before selling it to the British company Cable & Wireless, MCI "yanked anyone good out of internetMCI and put them into a new group called Advanced Services so they couldn't be taken by C&W," says one former employee. "Then they replaced those people with deadwood from long distance voice. C&W eventually sued WorldCom for screwing them, and it eventually got settled."

Big, Fat, and Sick

The buy-and-grow strategy that worked so well through the 1990s was to become WorldCom's nemesis. By the mid-1990s, WorldCom (which dropped the MCI name a few months after the merger closed in 1998) had acquired too many network assets. It had bought IDB WorldCom's network, and it had acquired 10,000 miles of additional fiber when it bought WilTel, MFS's network, UUNet, and eventually MCI. This forced the company to ignore its core business of long distance voice and get into the business of selling capacity to other telecom companies that didn't want to build their own networks.

Jack Grubman told *Red Herring* in 1997: "WorldCom is at the intersection of everything we like—no carrier in the world can offer the integrated set of facilities that it does. The company has nothing to lose and everything to gain."[15] He repeatedly encouraged investors to buy WorldCom stock if they didn't want to risk missing out on the best play in telecom history. At the time, his advice was welcomed by mutual fund managers and small investors looking to boost returns on their portfolios.

Meanwhile, the 1996 Telecommunications Act had brought new players into the market, and that boosted demand for these network services. In its 1997 annual report, WorldCom boasted, "WorldCom's fiber-optic networks are being built for the broadband data applications that will typify telecommunications in the next millennium."[16] The company said in its annual report that its Internet revenues had doubled in 1997 to $566 million, and its domestic data networking business was up by 35 percent to $1.575 billion.

George Gilder—writer, industry analyst, and effectively the spiritual leader for most broadbandits—was impressed by Ebbers and wrote: "He has shown the magic of entrepreneurial vision and temerity. Ebbers' fiber and Internet empire stands ready to release many more trillions of dollars in wealth and Internet commerce and communications, and threaten monopolies around the globe."[17]

What Ebbers didn't know was that he was the biggest fool in telecomville—no one was making meaningful money from bandwidth and data networks, because they were too expensive to build, manage, and upgrade. Competition was heating up from new companies like Qwest. Still, you couldn't fault Bernie for feeling smug—he thought he had finally cornered the market on bandwidth, which apparently was going to be a scarce commodity. After all, Sidgmore and O'Dell had been constantly telling him and the world that the Internet traffic was doubling every 100 days.

WorldCom and MCI had a large amount of combined debt. It was time to squeeze every last penny out of MCI. Ebbers asked MCI executives to stay in budget motels and fly coach. He got rid of the company cars. Sullivan was his hatchet guy, and he was glad to oblige. Having hitched his star to Ebbers, Sullivan rose to the top very quickly, the MCI acquisition coming as his crowning moment at the ripe old age of 36.

Oswego magazine called Sullivan "master of the mega merger." "Scott knows his numbers like no one else," Ebbers told the *Wall Street Journal* in 1998. "I don't think WorldCom would be where it is today without him."

CFO Magazine named Sullivan CFO of the year in 1998. (Ironically, a year later, it put Enron crook Andrew Fastow on the cover!) This was turning out to be a fantastic year for Sullivan. He bought a 4.11-acre plot for $2.45 million in Boca Raton, Florida where he was planning to build the castle of his dreams, but meanwhile he commuted in a private jet to Mississippi from his temporary 3,000-square-foot home in Boca Raton.

UUs versus Them

UUNet insiders had a certain arrogance. Sidgmore's gang never really thought much of the crew from the South and ran UUNet as a private club. The newest inductees to this club, which included O'Dell, were Vint Cerf, the "official father of the Internet," and Fred Briggs, the WorldCom chief technology officer (CTO). In 1998 WorldCom built a new facility in Ashburn, Virginia. The 535-acre campus was a state-of-the-art facility, with features that would have been at home in Silicon Valley. It had a 2.5-mile bike trail, a dry cleaner, a day care facility, a bank, and a gym.

The big man on campus was Mike O'Dell, chief scientist and vice president at UUNet. Notably, Sidgmore and O'Dell continued to tout their UUNet affiliation, not their affiliation with WorldCom. O'Dell had joined UUNet in 1987 as its 31st employee. He was short on people skills, but his technical brilliance made him a favorite with the nerds who were the real force at this particular facility. O'Dell's office was full of nerd gear like Ren and Stimpy dolls, and to relax he would organize balloon races on campus. He could get away with it, for UUNet was bringing in hundreds of millions in revenues for WorldCom in 1999. " 'If you're not scared, then you just don't understand' was a classic O'Dell aphorism, and he had many," said Ahmar Abbas, who worked with the company in various capacities including sales. "If you hadn't heard this a couple of times during the interviewing process, you would most certainly be parroting it by the end of your first week at UUNet. Everybody up and down the company hierarchy believed this as if it were religion."

This was an Internet operation and was being run as such. Folks were allowed their idiosyncrasies but still had to adhere to rules such as wearing khakis. Public display of underarm hair was not allowed.[18] "It's a matter of pride to [employees] that they're part of an Internet company," said Sidgmore. "I said from the beginning that as long as I'm here, we're not going to integrate it."[19]

"Sidgmore, who became vice chairman of WorldCom, had a cult following and was a pretty straightforward guy," said Abbas. But "no one wanted to stand up against O'Dell. He was a big huge guy and it was scary to be in a meeting with him." There was an inherent superiority among the UUNet employees, "a feeling of invincibility. UUNet folks thought that they were special, and many fell for that."

Expertise, Expertise Everywhere, and Not a Drop to Drink

WorldCom's acquisition of MCI and UUNet brought the company a stellar team, but UUNet folks scorned their MCI counterparts. Sidgmore and O'Dell wanted to keep a tight control on the broadband business. The only person who was spared from scorn from UUNet staff was Vint Cerf, because he was too well known and powerful in his own right. "It was an amalgamation of different companies. The atmosphere was too political, and turf battles were going on all the time," recalled MCI-alum Bynum. As a result, people were too busy pleasing their individual bosses to worry about the long-term fate of WorldCom. Brownnosing apparently was in the employee handbook, and people would bend over backwards in order to please seniors like Fred Briggs.

For instance, if CTO Fred Briggs wondered aloud whether World-Com should stop buying class five switches, which help guide your phone calls, and replace them with software that does the same (also known as soft switches), a whole team of executives would spend months trying to come up with a way to answer Briggs' question in the affirmative. "If you told someone that the cost was going to be prohibitive, the answer would be 'Fred Briggs wants it this way'," recalled one executive.

Executives would arbitrarily tell Briggs that it cost $400 per T-1 connection, when in fact the real cost of a T-1 was much more. Briggs, in turn, would parrot those numbers to analysts, who would then base their earnings and revenue estimates on the data provided by the company.

This created a chain of escalating lies. Former executives confessed that they spent most of their time trying to figure a way out of this mess, so as to not make Fred Briggs look like a liar. One trick would be to bill the customer for the amount of bandwidth they used rather than for a T-1 connection. Another trick, as would be revealed later, would be to exclude from the final bill the costs of equipment and operating those high-speed lines.

"I was selling for UUNet in Asia, and to me it was clear that the costs were higher than the sales prices," recalls Abbas. "But in those times nobody at the company cared about profit margins—you got compensated on revenue, so you only cared about revenue." The company was buying T-1 connections from other incumbent phone companies and then selling them at a lower price later. For instance, in Japan, it cost UUNet $2,000 per month to buy a T-1 connection from the local phone company, but Worldcom would resell it to its end customers for $1,500 a month. "We were selling below cost, so we assumed the other divisions were making money—after all, we were a very small portion of a very big company," recalls Abbas.

Many insiders believe that the result of all the acquisitions was that no one really knew what was going on inside the company. A decade-long buying binge was beginning to catch up with WorldCom. While it had a decent back-end system to manage the networks and get phone circuits installed, WorldCom didn't have the front-end system that could send bills to customers. Insiders have revealed that even as the company kept buying other carriers, there was little effort made to integrate the back-office systems. WorldCom's salespeople were using one system, while MCI salespeople were using different software. It created chaos and, as a result, running day-to-day operations was a lot of guesswork. Forecasting demand was even tougher. "In reality we could not get the politics of the two entities [MCI and WorldCom] to align," added Bynum. "It was just impossible to work. For example, it would take only four hours to provision a cross-country DS-3 connection, but it took more than a month to get the account started." Simple things like credit verification would take weeks, and getting the connections to work would take months, many insiders lamented.

According to insiders, the chaos inside WorldCom grew so acute that many salespeople were competing against one another. Thanks to one of its acquisitions, WorldCom ended up acquiring telecom rights to seven buildings in Boston. Tenants (mostly corporations) in those buildings who were eager to get Internet access signed up with WorldCom. The salespeople would go out and make a sale, come back, and fill out the paperwork. WorldCom in turn would then place orders for T-1 connections with the local phone company, in this case Verizon (née Bell Atlantic). Insanely enough, someone from Verizon would call another department at WorldCom, buy T-1 connections from WorldCom at a lower price (since WorldCom had prebuilt connections into those buildings), and then resell those connections to WorldCom at a higher price.

No one could guess, however, that things were slowly decaying inside the company. Pro forma revenues for 1998 were $30.4 billion, up 15 percent from 1997, with almost $2.2 billion coming from Internet services and $5.8 billion from other data services. In 1999, WorldCom reported sales of $37.8 billion and earnings of around $4 billion. At the time, no one had any reason to suspect that WorldCom had been lying and cooking the books—those revelations would come three years later, in 2002.

The executives did their part to hype the stock. "The Internet continues to grow at 1,000 percent a year in terms of bandwidth demand, and voice's need for bandwidth grows about 10 to 15 percent each year," said John Sidgmore, then WorldCom's vice chairman, to *Red Herring* in early 1999. "By 2004, voice will be less than 1 percent of all bandwidth."[20] No one dared question those numbers—it all seemed plausible because of the incredible growth in the number of dot-coms and other Internet companies around the world. WorldCom was one of the top five phone companies in the world.

"As long as UUNet could keep up the illusion of the exaggerated traffic growth, WorldCom was a darling of the stock market. The executives of WorldCom were only interested in keeping up the value of their stock offering, not whether there were actually any customers for the Internet bandwidth facilities that UUNet was deploying based on their story of exaggerated growth," Bynum lamented later.

By the end of 1998, WorldCom was the undisputed bandwidth leader, and its success (and stock price) inspired many imitators. That ultimately became a major headache for Bernie's crew. The industry was beginning to resemble an open-air market, and by late 1998 dozens of players like Qwest and Global Crossing had jumped into the bandwidth game. Ebbers had made a big bet on bandwidth, and it was going sour. And it was only going to get worse, as 1999 would prove to be a bad year for Ebbers and Sullivan. After absorbing MCI, WorldCom sold off MCI's SHL Systemhouse unit for $1.4 billion and indicated that it would spend that money to expand the WorldCom network. The company was running out of options—it needed to compete with the imitators such as Qwest and Global Crossing that were looking to eat Ebbers for lunch.

Managing with Blinders On

But while Rome was burning, the Neros were busy with their orgies. Money flowed like champagne at company events like the one in 1998, when about 60 to 150 people gathered for a long weekend at a castle a few miles away from Windsor Castle. It was a weekend to build company morale and bring together people from different companies acquired by WorldCom. One executive who was there talks about senior executives climbing poles and breaking expensive croquet sets. Things got out of hand when some executives decided to go to London in a cab and moon the pedestrians!

Many such events took place in exotic locales like Mount Fuji, Japan; Phuket, Thailand; and Langkawi, Malaysia. "It was a great time, and everything else will be anticlimactic from this point on," said one executive. Another reported that the Teatro Opera in Buenos Aires was rented for a private showing of *Les Miserables*, and a ballroom in São Paolo was rented for another WorldCom bash.

But the company didn't have the money to pay for much of this. According to various shareholder lawsuits, the company would resort to double-billing, slamming customers, deliberately understating costs, and delaying customer credits for billing mistakes as well as payments to vendors. The cash situation at WorldCom was getting worse, and the company's cost-cutting measures weren't always prudent. For instance, in 1999, WorldCom instructed its sales and marketing teams to stop buying research from research organizations such as Gartner Group and instead use information from research reports written by stock market analysts. It was a dumb move, because most corporations need market research to predict their competitors' strategies and figure out what new product offerings could be hot sellers.

Chaos being the order of the day, it was evident to many insiders that the wheels were coming off the WorldCom express. Because of an influx of new competitors, the prices of bandwidth had been in a swoon and costs were simply going out of control. "Whenever I had discussions with planning and marketing [departments] they would hem and haw, complaining of cash flow problems. Everything was being done on credit, not cash," recalled a senior company executive.

It was around this time that WorldCom started signing long-term, fixed-rate lines leases to connect its network with the networks of companies such as Verizon and SBC Communications.[21] While WorldCom owned a lot of capacity, what it lacked, like most independent phone companies, was the connection to the customer's office—the so-called "last mile" link. That connection was, and to a large extent still remains, a monopoly of the local phone companies. It also bought fiber capacity to fill out the holes in its network, even though the company had a fairly large network itself. The company signed long-term contracts with other carriers, some extending to 20 years.

The idea was that by buying all these leases, WorldCom could easily provision services—that is, provide voice and broadband connections—to any customer who so desired. It was the logical thing to do—the faster you offer service, the faster you can start generating revenue. However, Sullivan and others at WorldCom totally overlooked the competition from other carriers such as Qwest and Global Crossing and even AT&T.

In other words, they spent too much money on equipment and contracts and could not generate enough cash. In 2000, WorldCom had $39.1 billion in revenue and $24.8 billion in debt. The company had been borrowing like there was no tomorrow and spending it like a drunken sailor.

Stock Addicts Anonymous

All through the acquisitions binge, WorldCom would strip the acquired companies and salt away some money for a rainy day. For some time, Ebbers and Sullivan used these seeming reserves to make up for the short fall.[22] This gave the perception that the company was growing at warp speed, and that kept the stock flying high. Bernie had made a Faustian bargain with the devil named Wall Street. WorldCom's hypergrowth attracted big investors to the company, who spent billions buying World-Com stock. Being a growth stock, WorldCom had to show revenues that were always growing. It could not afford to show a slowdown in revenues, because that would drive down the stock.

Bernie's friend and confidante, Jack Grubman of Salomon Brothers, was doing his job by keeping his "Buy" rating on the stock. Like other high-flying Internet and broadband stocks, WorldCom's stock at the time was being valued on per-share revenues and EBITDA—earnings before interest, taxes, depreciation, and amortization. In reality, it was more like earnings boosted by irregularities, tampering, and dubious accounting. Still the stock continued to soar. It eventually hit an all-time high of $64.50 a share in June 1999, courtesy of Grubman,[23] who was one of the pivotal figures in the broadbanditry—he introduced Qwest CEO Joe Nacchio to Qwest founder and billionaire Phil Anschutz; he also advised Global Crossing founder Gary Winnick and, of course, was the informal sidekick to Bernie Ebbers.

Bernie and Jack were very close. Grubman even attended Ebbers' second wedding in Jackson, Mississippi, in 1999, racking up a bill of around $1,100 and then expensing it to his employer, Salomon Smith Barney.[24] He had been championing WorldCom for a long time, and until about July 1999 (when concerns about an industrywide slowdown in long distance voice revenues pummeled the stocks of companies such as WorldCom and AT&T), the ever-rising stock made him the rock star analyst everyone else wanted to emulate.

Grubman often attended WorldCom board meetings[25] and used to tell the company in advance what questions he would ask during quarterly conference calls with analysts and press. This way, the two could present the company in the best possible light. On August 20, 1999, Grubman released a report calling on investors to buy WorldCom and, in fact, "load up the truck!" He predicted WorldCom would double its earnings every two or three years through the first decade of the 21st century. This gave the stock a little nudge, but even he had to know that something drastic needed to be done because the company, with its expensive bandwidth leases, inept fiscal management, and inefficient processes, was running out of cash fast.

So Grubman advised Bernie: Buy Sprint, the third-largest long distance company in the world. It had a lot of revenue and lots of cash. The company even had a cell phone business—Sullivan could create a wealth machine with his accounting machinations. In October 1999, Sprint, which is based in Kansas City, became the center of a takeover war between BellSouth, a Baby Bell from Atlanta, and WorldCom. The initial offers from both companies were about $100 billion, but with the help of its friendly Salomon Bankers, WorldCom was able to up its offer to $129 billion—$115 billion in equity and $14 billion in assumed debt. The deal was announced with much fanfare, and MCI WorldCom's stock predictably surged.

On October 5, 1999, Bernie boasted on the CNBC cable network, "How in the world would we have been able to provide a competitive broadband strategy to a consumer, which is really the focus of this thing, other than through a combination of our companies where we have enough scale and scope."[26]

This distraction, if anything, was enough for WorldCom to raise more money from the markets—about $3.2 billion in debt in 2000. The noise around the merger also allowed Grubman to change the metrics with which he valued WorldCom. He would change these often, and many overlooked the changes. Jack used forceful phraseology and expressed such apparent conviction that investors forgave the sloppiness behind his ratings and price targets. Just as magicians' scantily clad beauties distract attention from the sleight of hand, Grubman's words distracted readers from his lack of rigor and objectivity.

A review of several of Grubman's recommendations and price targets in his published reports shows a seesaw nature in his ratings and in his ever-changing price target methodology. The price targets were often a "stretch," and individual investors were more susceptible to the practice. Sophisticated institutional investors often care little about the valuation methodology used by sell-side analysts, other than for the expected impact on the stock if the analysts have to change their rating or price target. Institutional buyers are highly skilled and well paid to do valuation work themselves. To the less sophisticated retail investor, though, the valuation methodology and results can be the deciding factors in whether or not to buy the stock.

Grubman referred to WorldCom's P/E-to-growth rate to justify his

$130 ($87 split-adjusted) price target in May 1999.[27] Then, 14 months later, he once again urged investors to purchase their shares, arguing that the $87 target implied a firm value-to-EBITDA ratio below that of his five-year projected EBITDA growth rate.[28] Grubman maintained his $87 price target until November 1, 2000, when, with little explanation, he slashed the price target to $45 (the stock was approximately $18.94). He came up with a new approach, and it was "based on" a 1.5 to 2.0 multiple of the five-year cash EPS compound annual growth rate of 17 percent.[29]

With stagnating revenues and increased capital expenditures that threatened to cut into WorldCom's cash flow, it was hard to justify his previous valuation of WorldCom. In the past, acquisitions always helped give an impression that revenues and cash flow were rising. And if the valuation dropped, the Sprint deal would fall apart and Salomon would miss out on a big bounty in advising fees. In addition, any slide in valuation meant the whole house of cards called WorldCom would go tumbling down.

Now, as an independent analyst, Grubman should have alerted the investors to the stalling revenues and increasing expenses. But since Grubman's loyalties were with Ebbers and Salomon Brothers, he gave the investors the shaft and again changed his valuation metrics. The new sleight of hand was called cash earnings. Cash earnings are derived by subtracting cash expenses from cash revenues. This differs from earnings in that it does not include noncash expenses such as depreciation.

But no one noticed this financial sleight of hand, because the telecom universe was focused on this gigantic merger. When the merger eventually fell apart, it wasn't because of Grubman's flawed valuation method. Instead, Ebbers, in his arrogance, had managed to piss off regulators who needed to approve the deal, first in Europe and then in the United States. As a result, the regulators created enough roadblocks to block the merger. In June 2000, eight months after the deal was first announced, the word came down: The WorldCom-Sprint marriage was off. "I think that was the end of WorldCom and Bernie," said Brian Thompson, a former MCI executive who had started and sold LCC International, another long distance company, by then. With no cash to feed the WorldCom monster, the whole Ponzi scheme was going to break down—it was only a matter of time.

Project Save Bernie

While WorldCom was publicly making a bid for Sprint, Ebbers was try-ing to secure his own future. In August 1999 Ebbers secured a loan of about $499 million from the Travelers Insurance Company, a Citigroup subsidiary and the former parent of Salomon. (Salomon was bought in 1997 by Travelers, which merged it with the brokerage firm of Smith Barney to form Salomon Smith Barney, a division of Travelers, which then merged with Citibank.)

Rick Olson, a Salomon broker based in Los Angeles, had been in touch with Ebbers and worked to arrange the loan, which was made out of Travelers' Chicago office to Joshua Timberlands LLC, a company owned by Ebbers. The loan was collateralized by Bernie's WorldCom stock holdings. This private company, based in Brookhaven, Mississippi, purchased about 460,000 acres of property in Alabama, Mississippi, and Tennessee for $397 million. Despite his aw-shucks "I'm a simple south-ern boy" act, Ebbers was simply protecting himself. In February 2000, Travelers loaned Ebbers another $180 million, bringing the total to a whopping $679 million. In addition to these loans from Travelers, Ebbers took out more personal loans. In total he had about $900 million in loans that were secured by WorldCom stock.

What did he need this money for? For buying timberlands in the American South, ranches up in Canada, and a boatyard, along with $800 million worth of WorldCom stock on margin. He also bought Angelina Plantation, a 21,000-acre property in Monterey, Louisiana, in 1998 and other farmlands that actually received a $4 million subsidy from the U.S. Department of Agriculture.[30] It was the stock purchases on margin that got Ebbers into trouble later.

Because many of Ebbers' personal loans were collateralized by World-Com stock, it was important that the stock stay above a certain level. As long as the value of Ebbers' WorldCom stock stayed above $900 million, all would be fine. If it drifted below that, Ebbers would have to make up the shortfall, known as the margin call.

The failure to merge with Sprint had taken some of the sheen off the WorldCom stock, which, by the fall of 2000, was down 71 percent from the stock's high in June 1999. Through 2000 and 2001, the WorldCom stock kept sliding. Grubman kept saying buy, but by now, everyone on

Wall Street knew there was no point paying much attention to Grubman. The stock, which had been at $53.06 a share at the end of 1999, slid down to $14.08 per share by the end of 2001.

The situation became so bad that during 2000 and 2001, WorldCom loaned Ebbers around $400 million to cover margin calls on his loans. These loans were offered at a rate of about 2.15 percent, far below the prevailing prime bank lending rate of around 4.75 percent. Ironically, instead of using all the money for meeting his margin calls, Bernie used about $27 million for personal reasons, including a $1.8 million payment for the construction of his new house. In effect, he was getting lower-than-market rates not only to save his skin, but also to live it up. This is in addition to the more than $77 million in cash and benefits that Ebbers got from WorldCom between January 1, 1999, and December 31, 2001—during which time shareholders lost in excess of $140 billion.

When asked about these loans, the company would say that if Ebbers sold stock on the open market, it would further depress the stock price, which would not be good for WorldCom investors. In reality, the only person they cared about was Ebbers, because it was Ebbers who had made everyone who mattered—in this case, the board of directors—extremely rich.

To be kind, WorldCom's board was under Ebbers' control, even though some have said that they were outright incompetent and toothless. Ebbers had substantial influence over the board's decision-making process and actions. Anyone with questions would be put in their place and, as would later be revealed, "critical questioning was discouraged."[31] The board's compensation and stock option committee was pretty much in Ebbers' hands, and they basically did whatever Ebbers asked them to do.

The key figure on the compensation committee was Stiles A. Kellett Jr. He was the rubber stamper of Ebbers' decisions. In exchange for letting Bernie run amok, Kellett got to rent a luxury Falcon 20F-5 jet owned by WorldCom for $1 a month. While he paid the sundry expenses, he did not pay the $1.4 million or so that the lease for the plane would cost on the open market.

Others on the board were equally compromised. Since May 1999, they had been receiving stock options for being on the board of directors and were getting extremely rich. Kellett, for instance, had about 1.2 million WorldCom shares, while others had more than 100,000 shares each. In total, the board of directors and the senior executives of the company, including Ebbers, held a total of 40.4 million shares. The share numbers were high enough to create a serious conflict of interest! In reality, everything WorldCom did was to save Bernie and his bad bets. And that included the accounting shenanigans that would eventually bring WorldCom down.

1-800-CONNED

Over the years, WorldCom's modus operandi was quite simple: Buy companies, add their revenues to WorldCom's revenue stream, strip their assets, and take charges worth billions of dollars to account for the costs supposedly incurred in connection with the merger. Now, these techniques might be overlooked if a company bought a rival or two. But by the end of 1998, WorldCom had bought about 60 companies for a total of $70 billion.

In addition, WorldCom was using other dirty financial tricks. It would include in the charge the cost of the acquired company's expenses expected in the future quarters. Instead of amortizing the cost of acquisitions over several quarters, it took huge charges in a single quarter, which made the company's financials even more confusing. That way, World-Com did not have to record these expenses in the period they were actually incurred. This helped inflate WorldCom's earnings incorrectly in future periods while deflating earnings in the period of acquisition. But investors chose to ignore that, as they assumed it was part of merger-related costs. This way, WorldCom could use the excuse of a merger to obfuscate their numbers and be assured of good future earnings. Of course, Sullivan would put something away for the rainy day in his cookie jar—accounting geeks call it merger reserves.

But by the end of 1999, even these dirty tricks could not keep up the appearances. Scott Sullivan, the accounting wizard and *CFO Magazine*'s CFO of the year, basically forgot the principles of Accounting 101 in order to save Ebbers. Sullivan was in charge of the WorldCom Finance Group. Being a close confidante of Ebbers and the de facto number two, Sullivan was the man where the buck literally stopped.

From 1999 on, Sullivan had been perturbed by the rapid erosion of

WorldCom's revenues as the company came under tough competition from its rivals. He dipped into his cookie jar of reserves, but that was not enough. WorldCom tried to buy Sprint, and that failed. The dot-com bust and degrading business were proving to be near-lethal body blows for WorldCom. Perhaps it was sometime in July 2000 when Sullivan decided to bend the accounting rules a little. He, along with sidekick David Myers, decided that the costs, or expenses, were getting out of control, and it was time to hide them.

But before he did anything, in August 2000, Sullivan sold stock worth $18 million. In October 2000, the company prepared for an earnings call, in which it would announce that it would miss earnings expectations. Inside the company, executives worried that Wall Street would now focus on WorldCom operations, now that the Sprint merger was dead. In an e-mail dated October 21, 2000, Sullivan told Sidgmore that the company was in a "really scary"[32] situation. Apparently, revenue from America Online, one of WorldCom's major clients, was up only 1 percent, mostly because the bandwidth price had fallen sharply and the Internet traffic growth had slowed. "Wow! I had no idea that the revenue growth had deteriorated that much," Sidgmore wrote back, adding that "it's going to take some pretty fancy explaining."[33]

Sullivan is the kind of guy who reminds you of this joke: A CEO looking for a new accountant asks all applicants what two plus two equals. Most answer four and don't get the job. The winner says, "Well, what do you have in mind?" Sullivan pretty much told Sidgmore that he would make some accounting changes, and that would mean better margins for some parts of the business.[34]

What Sullivan did not say was that he had it all figured out. What Sullivan was doing was capitalizing expenses. Capitalizing, according to accounting rules, is saying that the amount you paid for an item is not an expense but rather an asset of equal value to its cost. So instead of deducting it from the revenues as an expense, the company can depreciate its value over a period of time.

This is a common method of accounting for equipment like computers and office equipment, which do represent real value and whose value does depreciate over a period of time. But in Sullivan's case, he started treating regular expenses, such as those expensive leased lines mentioned earlier, as depreciable assets, or things that can be written off over a period of time. It was like going to a deli, buying $10 worth of baloney, paying for it in cash, and then coming home and saying, "Oh well, I think the baloney is an asset, and I am going to depreciate it one dollar a year over the next ten years." Never mind that the baloney will go bad in the refrigerator after a week.

For WorldCom, this scheme would create an impression of less expenditure, or, if you look at it from the other side, higher earnings. That's what Sullivan did, which netted him a $10 million special bonus from his boss in 2000. According to the *New York Post*, it seems that in 1996 America Online pulled a similar stunt but got off with a slap on the wrist. WorldCom's situation, however, was worse: At last count, the company had overstated its earnings by $9 billion.

In 2001, while Sullivan was scamming the investors, some of WorldCom's employees decided to screw the company for their own benefit. A few salespeople had figured out that since the acquisition of MCI, WorldCom had two major billing systems—one from WorldCom and one used by MCI. By switching existing customers from one billing system to another, the employees got extra commissions because the new entries into a different billing system were treated as a new account.

All of this would have gone on peacefully had it not been for WorldCom's falling stock and the worsening fundamentals of the telecom industry. By late 2001, the stock was slowly skidding toward $12 a share. On December 31, 2001, with its stock down to about $10, WorldCom's moment of truth was around the corner. It had $35.2 billion in revenues and $30.2 billion in debt.

If WorldCom's rise was spectacular, its downfall was even faster. In January 2002, Global Crossing went bankrupt. The ensuing brouhaha prompted investors, media, and analysts to look into other big broadband plays more critically—and when they looked at WorldCom, they didn't like what they saw. The media began asking questions about the loans to Ebbers, and in March 2002, WorldCom announced that the SEC was looking into accounting issues and loans to some officers and directors. On April 3, 2002, the company announced that it would reduce its work force. And on April 30, 2002, Bernie was shown the door, and Sidgmore was appointed the new CEO. But even Sidgmore could not save WorldCom—an internal audit revealed an accounting scam of gigantic proportions.

One of the first things Sidgmore did upon becoming the new CEO, however, was to reinstate free coffee for employees.

—Original Message—

From: John Sidgmore [mailto:no_replies_please@wcom.com]
Sent: Thursday, May 23, 2002 7:55 PM
Subject: Coffee Service Returns
Back, by popular demand . . .

As you know, earlier this year the Company-subsidized coffee service for employees was eliminated. I am pleased to announce that we are reinstating this service, effective immediately.

Over the next few weeks, Facilities (in the major sites) and local management teams will coordinate this effort throughout the Company.

John

—

WorldCom, Inc. Corporate Employee Communications

As Bernie taketh away, so John giveth. Nevertheless, the death spiral had begun. The company's credit rating was downgraded by rating agencies. On June 25, 2002, the company announced that it had overstated its earnings before interest, taxes, depreciation, and amortization for all of 2001 and for the first quarter of 2002. In July 2002, WorldCom went belly-up. It became the largest corporate bankruptcy in American history.

On August 1, 2002, Sullivan—who between 1995 and 2000 had sold WorldCom shares worth $45 million—was arrested on fraud charges. Too bad he would not get to live in his Florida Xanadu. Investigators are still trying to find the dirt on Ebbers, but it's proven difficult, since the man only used the phone or handwritten faxes, and his secretary answered all his e-mail. His computing skills, like his management skills, were not as good as his aptitude for screwing the shareholders. Other executives from WorldCom are also headed to the big house up the river. Some are willing to squeal in order to save their skins. Too bad they didn't think of the shareholders and investors soon enough.

In November 2002, WorldCom's life began a new chapter: The company hired Michael Capellas, the man who had sold Compaq Computer to Hewlett-Packard, as its new CEO.

2

ROCKY
MOUNTAIN HIGH

The lengthening shadows were the only indication of the passage of time, but the two men deep in conversation were oblivious to that. They had bigger things on their minds. Deep inside a dingy hangar at a suburban New Jersey airport, the two were debating the future of the telecommunications business.

Teterboro Airport, in the heart of Bergen County in New Jersey, is barely twelve miles from midtown Manhattan and on a clear summer day, one can catch a glimpse of the Empire State Building. The airport, one of the oldest on the East Coast, played host to American forces during the two World Wars. With such history providing the backdrop, the locale was fitting for the two men planning an assault on the world so far dominated by AT&T and the Baby Bells.

Phil Anschutz, a quiet billionaire, wanted Joe Nacchio to run Qwest, and he easily persuaded Nacchio that the job was the right fit. Anschutz was well aware of Nacchio's asking price—to be the top dog at a big telecommunications company. During their meeting, pleasantries were quickly disposed of and Nacchio and Phil got right to business. They discussed the future of telecommunications and why a national fiber-optic network was their ticket to riches. Nacchio reportedly told Anschutz, about what could be done with Qwest, "Even if you do them completely wrong, you'll still make a lot of money," he told Anschutz.[1]

On that cold November day in 1996, the duo reached an agreement and the meeting came to an end when they shook hands. Anschutz, then the eleventh richest man in the world, got back into his waiting executive jet and flew home to Denver, Colorado. Little did he know that this one handshake was going to bring him more media attention than he cared for and, along with his new comrade-in-arms, make him a member of an exclusive club, the Broadbandits.

Fourteen days later, Nacchio signed on the dotted line and got an $11 million sign-on bonus. The only thing still standing between Joe and the top job at Qwest was the board of directors. During a meeting with the board at a New York City hotel, "Joe basically scratched out a business plan on the back of an envelope and boom! We were off to Wall Street. He's the quickest study I've ever seen," said Robert Woodruff, then chief financial officer of the company.[2]

Nacchio's decision to leave stable AT&T for upstart telecom Qwest came as a shock. Joe was in charge of AT&T's $26 billion consumer long distance business and was one of telecom's rising stars. Qwest, however, was a sleepy old telecom contractor that, like others, had grand ambitions to become the king of the broadband world.

Nacchio's decision to join Qwest was announced with much fanfare on December 22, 1996. "There can be no more exciting challenge than to lead a company like Qwest. We have been handed a clean sheet of paper to complete the development of a company equipped with the latest technology in an industry that is transforming itself," Nacchio said.[3] With $11 million in the bank and more to come, it was going to be one merry Christmas. Within months of joining, he would help take Qwest public and drive it to the top of the broadband mountain before falling off the cliff—and, in the process, make Anschutz even richer.

The fiber frenzy had begun.

Mother Lode Redux

On May 6, 1859, John Gregory hit the mother lode when he found gold in North Clear Creek, just a few miles outside of Denver, stimulating a rush of prospectors who clamored for their own fortunes. Within two months, the area came to be known as "The Richest Square Mile on Earth".[4] A century and a quarter later, Philip Anschutz, a somewhat reclusive and shy oil baron turned entertainment magnate, hit a jackpot even bigger. His ticket to ride was a sleepy and struggling old railroad, the Southern Pacific.

Mr. Gregory's strike at what came to be known as California Gulch (despite being located in Colorado) yielded only $8 million in gold. That was a huge amount of money back then, enough to start a stampede of miners who threw caution to the winds and decided to go looking for gold in the snow-capped Rockies. The broadband gold rush was no different, for it brought in many imitators, but few profited as handsomely from the mania as the insiders at Qwest, especially Anschutz. According to published reports and regulatory filings, the 61-year-old billionaire earned more than $2 billion in the six years he owned Qwest.

An amazing capitalist, Anschutz has been described by *Fortune* magazine as the modern-day J. P. Morgan, largely because of his enormous holdings across many industries. His financial tentacles reach so far that one in every three Americans has come in touch with entities he owns or in which he has an economic interest. Anschutz achieved this by being a notorious perfectionist who first demonstrated his entrepreneurial flair in his 20s by fast-talking his way into owning some of the most lucrative oil-producing fields in Wyoming.

Today he is one of the largest owners of land in North America, with ranches in Texas, Colorado, and Wyoming. Chances are you have bit into peppers grown on his farm. He owns 30 percent of Shaq, Kobe, and company (the Los Angeles Lakers), five major league soccer teams including the New York/New Jersey MetroStars, and one of the National Hockey League's newer teams, the Los Angeles Kings. You might have watched movies at one of his theaters—he owns the United Artists, Regal, and Edwards theater chains, a total of 550 theaters and 5,738 screens. He is also building a 40 million square foot arena next to the Staples Center in Los Angeles (with buddy Rupert Murdoch), and if you watched the Oscars last year at the Kodak Center, well, Phil owns that, too.

Forbes magazine estimates his net worth at around $4.3 billion, making him the sixteenth richest American. Only 53 people in the world have more money than he does. In 2000, at the peak of fiber madness, thanks to his massive holdings in Qwest, he was worth $16 billion and, for a short while, was the sixth richest man on the planet.

He did it the old-fashioned way—by combining hard work, chutzpah, and some not-so-subtle arm-twisting. "For Philip Anschutz, life has been one long adventure in ferreting out hidden treasures in unexpected places," said *BusinessWeek* in December 1997.[5] Other media outlets, at least until recently, have been equally effusive in their writings on the billionaire. However, there is said to be a darker side to the man at the top of the Qwest mountain. To the world, Joe Nacchio was the chief executive of Qwest; however, many whisper that it was Phil, the chairman of the board, who was pulling the strings all along.

Anschutz is an "only in America"[6] kind of success story—he worked hard, was entrepreneurial, and knew all the right people. He counts Vice President Dick Cheney among his many friends and is a big supporter of the Republican Party. In 1986, along with Elizabeth Dole, Phil hosted a $1,000-per-couple fund-raiser for Ken Kramer, a Republican candidate for senator, aboard one of Anschutz's railroad cars.[7] His companies, his family, and he himself have contributed hundreds of thousands of dollars to Republicans.[8]

An avid sportsman, Anschutz works long hours and still hasn't lost his midwestern values, according to those who know him well. He still wears old jeans or old khakis with his favorite fishing jacket and a Timex watch. "He's a person of extraordinary integrity and a public-spirited individual. He's a very quiet, understated person and very intense all at the same time," said former senator Hank Brown, a Qwest director and president of the University of Northern Colorado.[9] Yet in May 2001, when Qwest was engaged in murky business shenanigans and the company's stock was sliding, Anschutz sold $408 million in stock. By fall 2002, he had reduced his holdings in Qwest from a whopping 86 percent to about 18 percent and had cashed in to the tune of $1.5 billion. And this did not include his sale of 16.7 million shares of Qwest to BellSouth.

At the same time, he is a devout Evangelical Presbyterian, known never to swear, and is "friendly and unpretentious, with an easy laugh and a good sense of humor."[10] He supports many charities, including The Foundation for a Better Life, which promotes basic values like honesty, hard work, courtesy, and gratitude.

Middle of Nowhere

Anschutz's story began in 1939 in a small town called Russell, Kansas. The town is on the map because of former senator and one-time presidential candidate Bob Dole, now a friend of Phil's. Son of Fred Anschutz, a successful cattleman but an erratic oil wildcatter, and Marian Pfister, Phil grew up with economic uncertainty as the family scrambled more often than not to make its mortgage. It is said that his grandfather, Carl Anschutz, had emigrated from Germany and founded a bank in Russell. Little Phil was still running around in his shorts when his family moved to Hays, another small town in Kansas, before they settled in Wichita, where he attended high school.

He was a mediocre student but worked hard, and soon he was attending college. As Phil was finishing up his business degree at the University of Kansas, his family faced financial problems, delaying Phil's plans to attend the University of Virginia's law school. Instead he joined the family oil-drilling business, Circle A Drilling, moved to Denver, and began his spectacular rise to the top.

The oil business is fraught with uncertainty, which young Phil learned quickly when his first venture almost went out of business before it even got off the ground, nearly bankrupting him. In the fall of 1968, 27-year-old Phil got a call from one of his rig supervisors with the news that they had hit a gusher in Gillette, Wyoming. Figuring he did not have any time to waste, Anschutz chartered a small plane and soon reached the oilfield. The oil was everywhere, and the whole area could explode any minute.

Most normal people (including oilmen) would have panicked. Not Phil, who got the well capped, and then went out and bought oil leases all around the region. He had a month to figure out how to pay for all that. He flew back to Denver, wondering which banker to call. At home, he turned on his television to see that his fields were on fire. "I tell you, I thought that was the end of me in business," he said in an oral history interview with the Colorado Historic Society.[11]

The story goes that Anschutz tried to convince the famed oil-fire fighter, Red Adair, to help him. Adair refused at first, but Phil badgered him until he agreed. In for a penny, in for a pound, young Phil must have thought. Then he convinced Universal Studios, which was in the process of making a film about Adair, to film the fire-fighting effort for $100,000. It was the single best deal he ever made. The film *Hellfighters* was a major hit for John Wayne, and Anschutz profited from the quenching of the fire instead of losing his shirt because of it.

The Wyoming episode demonstrates qualities that have made Anschutz a success: He is tenacious, patient, and stubborn—and always
willing to take a wager on the impossible. "I was pretty much left to sink
or swim on my own. You are very lucky if one out of 25 ventures hits
success," he has said.[12]

Boy, did he learn that the hard way! According to *Fortune* magazine,
there was a time when, as an oilman, he hit 30 dry holes at a stretch and
had banks cutting off his credit.[13] In the late 1970s, he learned about
some new digging technology and hit a gusher in northern Utah. That
strike was when life turned around for Phil. The strike turned out to be
the biggest oil field since Prudhoe Bay in Alaska. In 1982, he sold that
billion-barrel oil field to Mobil for $500 million, a year before the mar-
ket collapsed.

He quickly diversified into other areas, teamed up with other rich people like the Pritzker brothers (another reclusive family that owns the Hyatt chain of hotels), and in 1984 went after ITT, a large conglomerate
with disparate interests including some technology subsidiaries. Anschutz and the Pritzker brothers together assembled a 4.9 percent stake
in the company. When the conglomerate resisted, the partners backed
off, worried about negative publicity—Anschutz had always worked hard
to stay away from the media glare. Even today Anschutz jealously guards
the privacy of his three children—two daughters and a son—and his
wife, Nancy.

■ ■ ■

Anschutz's foray into the takeover world was short-lived, and in October
1984 he offered to buy Rio Grande Industries, Inc., parent of the struggling Denver & Rio Grande Western Railroad, for $496 million, or
about $50 a share. The company had sales of $385 million and earnings
of about $23 million in 1984. Only $90 million would come from Anschutz, but he would own 82 percent of the company. The rest came in
the form of debt from Wall Street. Phil was bargain hunting, and it was
clear that the Rio Grande shareholders were going to get the short end of
the stick, a pattern that would repeat itself later as Anschutz's wealth grew
and grew. In an interview with United Press International, Shearson
Lehman/American Express railroad analyst Mary DeSapio called the

$50-a-share bid "outrageous. . . . This is scrap value. Rio Grande officers might be good negotiators, but they are not negotiating for shareholders in this case," she lamented.[14] Nevertheless, the deal was consummated a month later.

Luck was not in Anschutz's favor, as he struggled with the railroad and had to worry about the merger between two of the area's much larger railroads, the Atchison, Topeka & Santa Fe and the Southern Pacific. Anschutz's team began lobbying hard with the Interstate Commerce Commission (which was later terminated), trying to block the merger, claiming that it would take away Rio Grande's access to Southern Pacific tracks.

Anschutz figured he could get out of this jam. He and the Pritzker brothers had attempted to buy Southern Pacific in 1983 but were rebuffed. Anschutz is known to be a history buff and a fan of Napoleon, and he often quotes Napoleon's famous line, "What do I care about circumstances? I create circumstances." The railroad tussle grabbed national attention. Rio Grande lined up support from some of the biggest American corporations, including US Steel. In the end, Anschutz won and, in a strange twist, ended up buying Southern Pacific in 1988 for about $1.8 billion. Did someone say tenacious?

"The combined railroads will operate under the banner of Southern Pacific," Anschutz said when announcing the merger on August 9, 1988.[15] "The Denver & Rio Grande Western will continue its own proud heritage of service as an integral and pivotal part of the overall system." The headquarters for the combined 15,000-mile, 15-state system was in San Francisco. The deal was done pretty much for free. In order to conclude the deal, Morgan Stanley and Company helped Anschutz issue $200 million in high-interest junk bonds and, in return, got a 25 percent stake in the railroad. The rest of the money was borrowed from other banks, making the combined railroad the fifth largest in the nation and one of the country's most leveraged companies.

Ironically, while Southern Pacific would make him rich beyond his wildest dreams, it was clear to all that his real interest in purchasing the ailing railroad was real estate. "These real estate assets have been, and will continue to be, analyzed in detail," Anschutz said then in a press release. "We are determining just what the railroad needs for present and future growth. What is surplus, we will endeavor to sell. But we will not dispose of property which is needed to support the railroad now or in the future."

A year later, Anschutz sold off $350 million worth of real estate, paying back the junk bonds that had been used to finance the acquisition. According to published estimates, Anschutz ended up selling nearly $2 billion worth of real estate, primarily to governments in Texas and California. Despite that, the acquisition was not going too well. Phil was taking it on the chin, and he needed to do something quickly.

A master of the deal, in 1990 he pulled the proverbial rabbit out of the hat when he sold 5 percent of the company to Japanese shipping giant Nippon Yusen Kaisha. This bought him time, and within three years, Anschutz managed to turn the corner and sold a 20 percent stake in Southern Pacific in a public stock offering in 1993. His net take was $45 million. The good news was dampened by his father Fred Anschutz's death that same year, at the age of 84.

In 1994, things began to look up for Phil and Southern Pacific. The railroad tapped the public markets twice, which gave Anschutz a chance to reduce his stake in the company to 32 percent. All told, his net proceeds thus far from Southern Pacific totaled $350 million. But it was clear he was tiring of the railroad game. A year later, he merged Southern Pacific with rival Union Pacific for $5.4 billion; his remaining 32 percent stock in Southern Pacific was now worth $1.5 billion.

Many believe that he got the better of Union Pacific, which overpaid. Based in Omaha, Nebraska, Union Pacific has many loyal supporters (including shareholders), and they resented overpaying for Southern Pacific. Walter Scott Jr., president of Peter Kiewit and Sons, apparently was one of them. This resentment would surface later when Scott would wrestle with Anschutz and Qwest.

Another reason for Omahans not liking Phil was SP Telecom, which was shrewdly kept out of the SP–Union Pacific merger. This little subsidiary was eventually destined to become Qwest.

Other People's Pipes

Kim Bottoms was used to waiting for her husband, Danny, who was somewhere out in the wilderness. Her only means of contact with her husband was a cellular phone, but it rarely worked, because Danny could be anywhere from the Rockies to the Louisiana marshlands to the baking high plains of New Mexico.

That was ironic, given that the 30-year-old former U.S. Army intelligence officer was helping build one of the most sophisticated telephone networks in the world. The Bottoms lived in Denver, Colorado, the new home base of SP Telecom, a network construction company Bottoms had joined in 1994 as the senior manager of network construction. He had come to SP Telecom from MCI, after accidentally drifting into the network construction business.

In 1991, Bottoms had been at loose ends. He had just left the Army and was trying to figure out what to do next. Born in Fort Leonard Wood, Missouri, Bottoms had earned a degree in chemical engineering at the University of Iowa before joining the Army for three years. It was sheer luck that he joined MCI, a fledgling phone company headed by Bill McGowan, a maverick entrepreneur who had vowed to break the AT&T monopoly on the phone business in the United States.

Lobbying hard in 1984, McGowan had succeeded in getting the U.S. government to deem AT&T a monopoly, a decision that led to Ma Bell's breakup into seven Baby Bells that offered local telephone service. AT&T had to content itself with selling long distance service. MCI, a scrappy start-up, had used aggressive marketing tactics and was literally taking dollars out of AT&T's pockets.

As its share grew, MCI needed bigger and faster networks to handle the growing traffic. In the early 1990s, MCI turned to SP Telecom, a small construction division of Southern Pacific Rail Corporation. Bottoms' job was to manage construction crews involved in the construction of the fiber-optic cable routes, and to make sure the networks were built on time.

Founded in 1988, the Southern Pacific subsidiary was essentially a construction company—think hard hats, plaid shirts, and Miller Lite. This was not the first time the railroad had built a national network— an earlier effort had become Sprint Communications, a long distance operator.

SP Telecom had developed a special plow, which, when mounted on a railcar, could rake through literally anything and could install fiber in difficult terrain with unheard of efficiency. Costing $1 million, the bright orange, 76-ton contraption, dubbed the Rail-Blazer, had two arms that could extend fifteen feet on each side of the rails. It took two locomotives to pull the special vehicle at a maximum speed of four miles an hour. From a distance it looked like a giant spider crawling over the flatlands. "We can move rocks the size of a Volkswagen. It's like pulling a plow with a 6,000-horsepower tractor," Dan O'Callaghan, SP Telecom's vice president of construction, told the *San Francisco Chronicle* at the time.[16]

As it moved, the Rail-Blazer plowed a four-foot-deep ditch on either side of the tracks and thus made way for laying down conduits. It could lay down six conduits at a time, and each conduit could hold up to 200 thin fibers, which would be used to carry voice traffic. At the time, each fiber could carry about 32,256 phone calls simultaneously. An excavator (just like a backhoe) mounted on another car would fill up the hole. A crew of about 250 would follow this rail-plow work train around. The men earned only $12 per hour, but the operation cost a whopping $125,000 per mile.

While it took about a year to link San Francisco with Los Angeles, it took two years to get San Jose and Denver connected. "We crossed the Sierras and the Rockies, probably two of the toughest construction projects undertaken in recent years," SP Telecom's president Doug Hanson told the railroad's then hometown newspaper, the *San Francisco Chronicle*.[17]

"It was back-breaking work," recalled Bottoms. "We would work 10 to 12 hours a day, six days a week." The biggest problem for SP Telecom was to get the railroad to cooperate. In order for the Rail-Blazer to move, SP Telecom had to get track time for itself, and most of the day the crew would just sit there, waiting for hours at a time to get clearance from the railroad dispatcher. And sometimes things got a bit hairy. While building a Denver–El Paso link, the company had managed to lay down conduits all the way to southern and northern Albuquerque. But smack in between the two points were five Native American lands, also known as pueblos, part of the Native American reservations. "They would not let us move," said Bottoms. The status quo would continue for almost a year, and it would take several million dollars to end the standoff.

The biggest test construction crews encountered was the Moffatt Tunnel, one of the longest tunnels in North America and the sixth longest in the world. Named after Denver railroad pioneer David Moffatt, the 6.2-mile tunnel bores through the Continental Divide, 9,239 feet above sea level. For four months, working only the two hours per day that trains were not passing through the tunnel, Bottoms' crew struggled to hang a steel pipe from the ceiling of the tunnel.

Still, it was good business for SP Telecom. Americans were so busy gabbing with each other that they were rapidly filling up the cables that carried their voices. SP Telecom reportedly generated almost $1 billion a year in revenue, although the privately held company that was mostly owned by The Anschutz Family Investment Company never publicly disclosed its sales. Phil Anschutz had hit the mother lode yet again.

The SP Telecom operation was a perfect example of Anschutz's business philosophy: It is easy to make a big fortune from a small fortune, especially if you get others to pay for it. Even though he invested some money in the independent SP Telecom, it was more for the operational needs of the company. The money that MCI, Frontier Communications, and WorldCom were paying for SP Telecom's construction services allowed the company to drop a few fiber lines for its own use as well. Except for the cost of fiber-optic cable, a few thousand dollars per mile, SP Telecom built its network for free. By the time the construction was finished in 1999, the network would span 18,500 miles and would connect 150 cities worldwide.

In 1995, SP Telecom bought a small telecom company in Dallas, Texas, named Qwest. Within weeks, SP Telecom changed its name and got into the business of selling long distance telephone services. It was now competing with some of its customers, like MCI and Frontier, for the telecom dollars. It was no surprise they got into this game, given the low costs Qwest had incurred in building this network. By getting into the long distance business, it could easily undercut others, and even if the company carved up a minuscule share of the total long distance business, it would be enough.

Anschutz, always a keen observer of trends, was quite aware of the pending telecom deregulation and decided to go for broke. It was time to get Joe Nacchio.

The Fiber Cowboy

"I bet I can get you Joe Nacchio," boasted Jack Grubman in a conversation with Anschutz in New York.[18] In the fall of 1996, Grubman, who worked as an analyst for Salomon Brothers, was viewed as Wall Street's foremost authority on telecom, and a man with a golden rolodex. His friends included Bernie Ebbers, chief executive of WorldCom. And Grubman delivered.

Grubman knew that Nacchio was the perfect man for the job. He was loud, brash, aggressive, and he wanted to win at any cost. If central casting were to call for an archetypical wise guy, chances are they would want somebody who resembles Nacchio. He liked driving Porsches. He wanted the media attention and the power that goes with being rich. On the first day of his job at Qwest, Nacchio received a pair of cowboy boots and a note from Anschutz that said: "Welcome to Denver. Horses and guns to follow!"[19] Anschutz later gave Nacchio a horse and saddle, and Joe named the steed Q (Qwest's symbol on the New York Stock Exchange). The urban cowboy could now pass for the real thing.

"We became a telecom company when Joe joined, and the construction business became the backwater of the business," said Bottoms. Most of the senior management felt the shift of power and decided to leave. Joe brought in a lot of his cronies from AT&T. "I had to leave, because I didn't know the secret AT&T handshake," said Bottoms, who went to work for Mastec, a telecom infrastructure company based in Miami.

At the time, Nacchio was well known in the business for his dynamism and his controversial character. He loved to get into corporate brawls and wasn't happy until he won. In some ways Nacchio was a modern-day, executive version of Napoleon: Both were short in stature but had giant-sized ambitions. Like the French warrior, Nacchio was part dreamer, part pragmatist, but mostly a street fighter. Upon joining Qwest Joe claimed that until then he had been "an entrepreneur struggling inside of a big company."[20]

Of course, his entrepreneurial leanings came at a price: In one interview with *Inc.* magazine, he lamented the loss of luxuries that had come with the big job at AT&T. "When you get into the backseat of your car and it doesn't go, because you don't have a driver anymore—the little things like that. Every time I have to drive into Manhattan to take a meeting instead of taking the helicopter, it really brings it home to me."[21] He justified his decision with the knowledge that he had been fighting a losing battle at AT&T. "I tell you, when you play the whole football game on your side of the 50-yard line, you generally lose the game," he said.[22]

In reality, while Nacchio was bitter at losing the chance to be the top guy at AT&T, he had come a long way from his humble blue-collar beginnings. Born on New York City's Staten Island, his father, a longshoreman turned bartender, had moved his family to Brooklyn when Joe was still a baby. As a teenager, Joe enrolled at Stuyvesant High School in Manhattan, which was known for being one of the toughest high schools in the country. An academy of future leaders, its alumni include former New York Governor Mario Cuomo and *Charlie's Angels* star Lucy Liu; its teaching staff included the likes of Frank McCourt, author of the Pulitzer Prize winning autobiography *Angela's Ashes.*

In one conversation, Joe recalled how he used to take a subway all the way to the Bronx to train for cross-country races. Once he finished the New York City marathon with a bleeding foot. "At six in the evening on my way back, I would do my math homework on the train. My day was all planned out," he added. Since Stuyvesant was chock-full of smart kids, Nacchio learned about competition. It was at Stuyvesant that Nacchio's competitive streak first emerged. He developed a 16-hour-a-day work ethic.

He went on to attend New York University on borrowed money, and in the process became the first in his family ever to go to college. But he lacked a post-college plan, and his entry into the world of telecommunications was all but an accident. In 1969, as part of the on-campus recruitment drive, Nacchio was hoping to get a job with consumer goods giant Procter & Gamble. Instead he accidentally walked into an AT&T interview and ended up working for the company for the next 26 years. Rising through the ranks, it was not until he was 35 when Joe started to dream big.

"AT&T sent me to MIT's Sloan School of Management, and there I was with the best and the brightest from America—and at that time I kind of realized that I was good enough to compete with them," he recalled. He came back from his New England sojourn a different man, and as days went by he became more critical of his employer.

He rose through the ranks, first running the highly profitable networking and corporate business for Ma Bell. In August 1993 Nacchio was promoted to president of AT&T's $24 billion-a-year consumer long distance division, with one mission: to stop MCI and Sprint from taking market share away from AT&T in the highly lucrative long distance business. This was the perfect podium for Joe to showcase his street-fighting skills. In the long distance wars of the early 1990s, folks at MCI (now part of WorldCom) learned the hard way what it was like competing with Nacchio.

AT&T management knew it needed someone with Nacchio's aggression to rescue Ma Bell's $24 billion-a-year, bread-and-butter business from the wolves. MCI had grabbed a 25 percent market share, and Sprint had nibbled up another 10 percent share of a market that AT&T had owned for nearly a century.

Putting Nacchio in charge of the business was perhaps an uncommonly astute decision by former AT&T chief executive Robert Allen. "Joe was the first manager in the Bell system who had the face and personality and charisma," recalled Jerry Taylor, the former head of MCI who was on the receiving end of Nacchio's salvos. "Prior to that, AT&T was legions of faceless people, and Joe was the first guy to step out and fight it out. So he was a target for us as well, and he was very predictable." The MCI–AT&T battle for long distance domination put the Pepsi versus Coke wars of the 1980s to shame. Both companies were pumping out nearly 50 different television spots a year, spending over $200 million each to out market each other.

While aggressive, many also remember Nacchio as a pragmatic leader. For example, when he figured that it was costing $300 each to get new customers signed on to AT&T, he came up with a killer idea: Give consumers $100 to switch over, and save $200 per new customer. It was his signature on the $100 dollar check that Ma Bell used to lure consumers from MCI. Some believe that was just the beginning of the end for the long distance business and criticize Joe for that move even now.

Thanks to his determination and energy, eventually AT&T made a comeback. But it was a long, bloody fight that put Nacchio on the telecom map. Even his adversaries were impressed. "Joe rallied the troops at AT&T and gave them a target to go after," said Taylor. While to the outside world it seemed that Nacchio was headed to the top, insiders at AT&T say that Nacchio did not have a chance. When then-CEO Robert E. Allen appointed John Walter, former head of R. R. Donnelley, to replace him in November 1996, Nacchio felt spurned. He clearly had an eye on the top job.

Even years later, the snub still bristled. "I was passed over because frankly I was more outspoken, more performance oriented, and at the end of the day I said what I believed in," said Nacchio during the course of a conversation in early 2000. What Nacchio doesn't say is that in all likelihood his Italian name and working-class background prevented him from taking part in AT&T's country-club culture. After that disappointment, no wonder that when Anschutz promised Nacchio the moon, the stars, and all the money in the world to move to Qwest on that cold November day, it didn't take much convincing for Nacchio to sign on.

Spurned by the senior management at AT&T for the top job, it seemed he yearned to destroy the company. A few months after he had taken over as the chief executive of Qwest, Nacchio was part of a team from Qwest that met with a group of editors and reporters at *Forbes* magazine's Fifth Avenue offices. There, he nibbled on a croissant and casually went on to outline his vision of the telecom world, how broadband would change everything. But the most memorable quote of the day was, "Who knows, one day we could even buy AT&T." The silence around the table was deafening, but what no one suspected was that Nacchio had a game plan to do it.

Pump Up the Volume

To realize this plan, Qwest would need stock—a lot of hyperinflated stock. And in order to obtain it, the company needed to go public quickly. Joe started working for Qwest in early 1997, when Qwest was not a telecom heavyweight but a sleepy, privately held contractor with 700 employees and $225 million in annual sales. Some describe the pace inside Qwest as frenetic after Nacchio arrived. The hustle was necessary, for the company was priming itself for a public offering and Nacchio wanted to make a lot of noise with it. Six months after he took over Qwest, in June 1997, the company sold 13.5 million shares at $22 a share, and raised a whopping $297 million. In the process Nacchio's paper worth jumped to $20 million.

Nacchio's timing was impeccable, at least in the beginning. With the trendsetting August 1995 public offering of Netscape Communications, a simple browser maker had woken the American masses to the Internet, which until then had been an academic curiosity. Internet access providers such as Earthlink, Pipeline (later PSINet), Netcom, Mind-spring, and thousands of others could not keep up with demand from curious Americans who wanted to get on the Net and send e-mail. America Online was growing so fast that its customers were having trouble logging on to the online service.

This was perfect for Qwest, which sold capacity to companies like Earthlink as well as to other phone companies. With consumer demand filling the Internet providers' pipes, it wasn't long before Qwest started running out of capacity. New technologies allowed Nacchio to make his pipes better, bigger, and faster, but this took a lot of money. Lucky for Qwest that Wall Street was giving out cash in buckets in the form of debt. Like a college student with his first credit card, Nacchio headed straight to the mall. Before long, the company had raised almost $2.3 billion in debt, mostly to finance the technical improvements on the network and keep the businesses going.

A deal with GTE Communications for 12 transcontinental fibers in 1996 had helped build out and expand Qwest's network, but it had also helped create yet another low-cost competitor to the three long distance players—WorldCom, AT&T, and Sprint. In 1997 and 1998, Qwest was one of the few new companies with its own network and was on a roll. "[Nacchio] allowed you to do whatever you need to get the job done. He rewarded those who delivered, and if you failed you were fired. If only you could stomach the pace," recalls an executive who worked at Qwest at the time. Nary a week would go by when Qwest would not announce a new customer for its fiber capacity.

Using Qwest's white-hot stock as currency, Nacchio went on a shopping spree. First on his shopping list were three small Internet service providers: Supernet, Phoenix Network, and EUnet International. In 1998 Qwest bought LCI International, then one of the largest long distance company in the United States, for $4.4 billion. At the time, it was a bold move for Qwest: Its own annual sales were $696 million, while LCI had revenue of $1.6 billion. One former executive who came into the Qwest fold via the LCI acquisition and spoke on the condition of anonymity recalled, "When Qwest bought us, LCI was a process-oriented company which was number three in the long distance business, but it wasn't going to take over the world. LCI never thought big enough. Joe came in thinking that we could take on AT&T and we could be number one, and since he had the network assets, he wanted to do that in six months. Joe had the vision and a whole new energy level."

Having learned his lessons in the fabled phone wars, Nacchio knew he had to undercut his rivals. Qwest offered long distance phone calls for 7.5 cents a minute, a move that pretty much knocked the wind out of the long distance business. The low prices basically took away all the profit margins from the business. AT&T has never really recovered from that blow. "This guy [Nacchio] had a gun to his head and he [Nacchio] was pulling the trigger," recalled Brian Thompson, the chief executive of LCI. "Even in 1998 it was clear that he did what he wanted to do."

"Sure, I knew back then that the horse they were riding on wasn't going to win, but the market was rewarding them with higher multiples, and we decided to sell out when the price was right," added Thompson. In hindsight he pointed out that since new technologies such as Dense Wavelength Division Multiplexing were being developed by companies like Ciena, the market was going to be flooded with bandwidth. This technology allows you to splice the beam of light into many colors and increase bandwidth over existing fiber-optic backbones without installing new fiber. It was going to make bandwidth a commodity, and that in turn would have killed companies like LCI. Thompson knew it was time to get out, and Nacchio was paying top dollar anyway.

As a battle-scarred veteran of long distance wars, even Nacchio had learned that long distance was not a defensible franchise, and he needed access to the homes and businesses—now called "the last mile." So he sold 10 percent of Qwest to BellSouth, a smaller Baby Bell that served states like Georgia and the Carolinas, for $3.5 billion in April 1999. The rationale was that BellSouth could help fill some of those empty pipes Qwest owned.

Meanwhile, Qwest revenue rocketed to $4 billion by the end of 1999, with $458 million in profits. Stock had surged from $7 (split adjusted) at the close of its first day of trading on the NASDAQ to $25 at the end of 1998. This was nothing compared to the triple-digit gains being posted on a regular basis by dot-coms that were going public at the time, but it was still a spectacular enough increase to make Qwest one of the early success stories of the broadband boom. The more analysts and pundits talked up high-speed connections using cable and DSL modems, the more Qwest's stock defied gravity.

Around this time, Qwest would decide to move into the Web-hosting and software-on-demand businesses—businesses that sounded really cool in a press release, but in reality were going to cost Qwest a ton of money and perhaps not generate much revenue.

Qwest was already spending billions on building its network, putting so much capacity into the ground that the company had no option but to undercut others in order to fill up its pipes. Some executives interviewed said that there was the belief inside the company that it could withstand any kind of pricing pressure and drive competitors into the ground. "We wanted to sell bandwidth at any price," said one executive who prefers to remain anonymous.

Another insider who was privy to the internal wrangling at Qwest recalls that the company was engaged in a game of one-upmanship with other broadband players. For instance, at one point, desperate to get business and grow sales, a senior executive at Qwest is said to have promised Microsoft that it would build an exclusive backbone for the software giant for a mere $300 million. Did this telecom equivalent of a used car salesman stop to think that Qwest had spent billions of dollars building a network that was running mostly on empty and couldn't afford to be making special cut-rate deals that would no doubt end up costing, rather than making, money? Still, it made a nice press release and in all likelihood kept the hype balloon from deflating.

A few months later, on December 14, 1998, Microsoft announced that it would invest $200 million in Qwest. An overjoyed Nacchio remarked, "There's an exploding market for complex Web-hosting software and systems. We needed people who understood the applications and the operating system, and there is no better company for that than Microsoft."[23] It was amazing to hear Qwest talk about the future, when businesses were having a tough time getting even 1.5-megabit T-1 connections installed. Qwest was building the equivalent of a 1967 Ford Thunderbird when most people were happy to be able to get a Model T.

In April 1999, Qwest decided it needed to go global, in what would turn out to be the company's third strategic mistake—the first two being cutting long distance prices too low and getting into the ephemeral Web-hosting and applications service provider business. The company established KPNQwest, a joint venture with KPN, the Dutch telecommunications company that was planning to flood Europe with fiber. This was going to be the bank-breaking move for Qwest, though at the time Nacchio was cheered by everyone from the media to the analyst community.

The venture, which expected to raise approximately $800 million of debt to complete the construction, was Qwest's first frontal attack on the global markets. Joe thought he had made a good call—analysts were telling him that the communication services in western Europe, the addressable market for the KPNQwest venture, would grow to approximately $224 billion by the year 2001, with five countries—Germany, France, the United Kingdom, Belgium, and the Netherlands—representing 65 percent of the total.

But, as we learned the hard way, analysts are often off the mark when it comes to making market forecasts. At its peak, KPNQwest generated about 60 percent of Internet traffic in Europe. Revenues never crossed $2 billion, and in the end it cost Qwest about $1 billion in much-needed cash. In the summer of 2002, KPNQwest declared bankruptcy. Still, KPNQwest had served its purpose: At its peak, Qwest's stake in the European venture was valued at around $8 billion, and that kept Qwest shares flying high on the stock market—long enough for insiders to cash out.

Nacchio's incredible run at Qwest would have come to an early end had he not opted to make a hostile bid for US West, the Baby Bell that served sparsely populated mountain states like Colorado. By buying and bulking up, Nacchio was stealing a page from the book authored by the ultimate broadbandit, Bernie Ebbers of WorldCom.

Ring This (Baby) Bell

Baseball season was getting started, and Joe had been looking to catch some Colorado Rockies games when he heard that US West executives wanted to meet with him. US West, the weakling among Baby Bells, was ready to throw in the towel. It was May 4, 1999, when US West allegedly proposed a merger with Qwest, even though Qwest and other parties involved have always denied it. Nacchio heard them out, and decided to pass. US West had been dubbed "US Worst" by Nacchio and Qwest executives, who had laughed at it for being a poorly run, bureaucratic mess. The details of the events are chronicled in *Forbes*.[24]

Ten days later, US West Chairman Sol Trujillo apparently met executives from Global Crossing, who jumped on the opportunity, and on May 17, 1999, the two companies announced a $35.5 billion merger. For Nacchio it was too much to bear. He was in the race for telecom domination, and the pesky little telecom neophyte and former junk bond salesman Gary Winnick was going to upstage him. Winnick was the founder and chairman of Global Crossing. "They're in play. Get out the book on US West," Nacchio told his senior management.[25]

Qwest had never pursued a hostile acquisition target. Behind the scenes, Nacchio picked Afshin Mohebbi (former president and chief operating officer of Qwest), who had only just started working for Qwest, to manage the takeover. The son of Iranian immigrants who fled to California during the politically tumultuous 1970s, Mohebbi is calm, unassuming, and detail-oriented. He finished high school by age 16 and college while still a teenager. A stint at Pacific Bell landed him a gig for the Brits, and by the time he was 34, he was named to British Telecom's management board. "Afshin is going to be an outstanding addition to our strong Qwest management team," said Nacchio at the time in a press release about Mohebbi's appointment.[26] "He brings energy, intellect, and a new perspective achieved over a series of fast-track assignments that will help us continue to grow our business and change the communications industry."[27]

While Nacchio got all the limelight, it was Mohebbi who kept the ship on course. As luck would have it, Mohebbi was not yet working for Qwest when he came to visit Denver on the day Global Crossing made a bid for US West. After the news broke, he ended up checking into the local Hyatt, buying clothes at the local mall, and staying in town. "This is the ride of a lifetime. From the minute I showed up, it was just hectic. It was a blur all the way from early May to September. I like the fast pace, anyway. But this was an unbelievable pace," he later told *The Denver Post*.[28]

In less than three weeks, Qwest launched a hostile bid that pegged US West's value at $40 billion, a 30 percent premium over the stock price. Qwest also wanted to grab another, smaller telecommunications company, Frontier Communications, which had been in merger talks with Global Crossing. The young were going to eat the old, for it was Frontier's millions that had helped Qwest grow in the early days.

Nacchio eventually won the US West battle, but his victory was not well received by Wall Street, which scoffed at the idea of merging a high-growth broadband company with a slow-growth Baby Bell and sent the Qwest stock tumbling. US West chairman and lifelong Bell head Sol Trujillo, who was said to still be in a snit over the early rejection, didn't really want to do a deal with Qwest and instead wanted to go with Global Crossing. It is said that last-minute intervention by Anschutz helped smooth things over and thus saved the merger. Qwest got US West; Global Crossing got to keep Frontier as a consolation prize.

In the end, US West and Qwest consummated a $36.5 billion deal dubbed a "strategic merger," even though it was clear to one and all that Qwest was taking over the century-old Baby Bell. Then an armistice was called and a new Office of the Chairman was formed, where Anschutz, Trujillo, and Nacchio were supposedly to work together to help guide the new company into the new millennium. The executive powers would be shared equally between Qwest and US West.

That did not happen. The personality clash between Nacchio and Trujillo was too much. Apparently, in late February 2000, Trujillo told Nacchio that he was going to resign once the merger closed. Nacchio could not keep his mouth shut, and the very next day reporters got wind of the impending departure. Trujillo announced his resignation, and Nacchio promptly reduced former US West managers to bit players in his new regime. Ironically, in the midst of this drama, the German phone company Deutsche Telekom, according to published reports, was offering $90 a share for Qwest, or about $100 billion for the combined companies. Nacchio wanted to sell the entire thing, but US West folks got their revenge and blocked the deal. Had he been able to sell the combined entity to the Germans, he would have looked like a genius, with many more millions in the bank.

As soon as the merger closed, Nacchio slashed dividends and moved from the ramshackle Qwest offices to the plush US West headquarters. When he got there, Nacchio discovered that the air conditioning would turn off at 4:30, and the elevators stopped running to the 52nd floor (where all the executive suites were located) at 6 P.M. To him, this clearly indicated a cushy corporate culture. US West senior executives would work eight hours, and that, to Nacchio, was anathema. He didn't like the "It's okay to be lazy" message that got sent when the AC shut off in what he considered to be the middle of the afternoon—and he quickly remedied that situation.

Those in the know believe that it was Phil Anschutz who orchestrated the US West bid. For years, it had been rumored that Phil was the real puppet master at Qwest. Nacchio admitted as much later. "Phil Anschutz and I were close friends for five and a half years. I spoke to Phil two to three times a week. Every major decision I made at this firm, I sought his counsel. In the old Qwest he was the majority owner," Nacchio said. "I always went to Phil Anschutz when I needed counsel. Many times I would get calls from Phil just to find out what was going on. Phil was very involved."[29]

Anschutz, it seems, kept Joe on a very short leash, and Nacchio in turn demanded almost slavish loyalty from his staff. A workaholic and tough taskmaster, he ran through executives like Kleenex. He surrounded himself with yes-men who would not dare question his wisdom. In his mind, it was clear: There was only one star of the Joe Nacchio Show—Joe Nacchio. Brian Thompson, the former LCI International chief executive who, like many others, did not stay at Qwest for long, questioned the wisdom of building such a huge network, and was quickly rebuffed.

"I think Phil [Anschutz] is a reasonably complex and calculating person. He assumed there was a telecom story, and Nacchio is not dumb," said Thompson. The compelling reason Anschutz would bet on Nacchio was because he was an "old industry guy who on the face of it knew the industry, and was from AT&T," Thompson pointed out.

Nacchio was driven by his ego, ambition, and his insecurities. The first from his family to attend college, he was always aware of his modest background. Perhaps it bothered him, and he has often said in press interviews that he was "not born into wealth."[30] In many ways he was justifying his success, both as a chief executive and as a person. Money became a scorecard at some point—he had to be better and bigger and richer than most others. Success and its requisite display were important for Joe. His ability to fly into Manhattan on an AT&T helicopter, or to fly back to New Jersey from Denver in a private jet on the weekends to spend time with his family, was a symbol of the success that came only with a lot of money.

Winning US West was also a symbol of success for Nacchio—he had beaten Gary Winnick, and thereby perhaps even decided both their

ROCKY MOUNTAIN HIGH 59

fates. In an interview he said that he knew he was doing the right thing by buying the Baby Bell, Wall Street be damned. What did he know at the time? Was he worried that Qwest was a house of cards, and that bandwidth prices already trending down were going to skid faster? Or was he worried that sooner or later the growth would end and the reality would catch up with his company? Time would eventually answer those questions.

Riding the Light to Nowhere

Because the merger with US West was so huge, it had to go through a regulatory process. The company at that time had a total debt of $19 billion. It needed huge amounts of cash just to service that debt—that is, to pay the interest on $19 billion in borrowings.

By mid-2000, the telecom and broadband markets were suffering from overcapacity. Many competitive local exchange carriers, who were Qwest's wholesale customers for things like Internet capacity, were headed to the bankruptcy courts. This was beginning to eat into Qwest's revenues. However, only a handful of people knew the business realities of Qwest.

Meanwhile, the company was using cheap tricks all throughout its operations. For instance, in an effort to goose revenues, the company changed the publishing schedule of its Yellow Pages directory. Normally the directory was published in January, but this time around, the company published it in December 2000—and promptly tacked all sales on as part of Qwest's 2000 revenues. In addition, Qwest was discounting the hell out of advertisements in the Yellow Pages to show growth. The sales at Qwest Yellow Pages were up 6 percent while those of BellSouth were down 8 percent.[31]

If that was not enough, the company indulged in something called flashing. First reported by Chris Byron, a columnist for *Red Herring* magazine, flashing is a practice where a telephone company books *projected* revenues from a customer, say, $100,000 in a quarter, even though in reality the customer spends only $20,000. Quoting a sales manager, Byron wrote that the manager "was ordered by a regional vice president to flash a $30,000 sale for $275,000 instead. The practice was common." Byron quoted one of his sources as saying, "The pressure was on because of unattainable forecasts by Nacchio."[32] But to the outside world it looked pretty good.

On Wall Street and in the popular press, Nacchio was being hailed as a hero and a visionary for having bought US West and for saving his company from going the way of the other broadband players. Through the second half of 2000, the Qwest stock stayed above $32 a share. Using creative merger accounting, the company managed to report good results in the first quarter of 2001, which pushed the stock to about $42 a share. This gave many insiders enough time to dump their stock. In the April-May 2001 time frame, Nacchio dumped 1.26 million shares worth $49 million.

By then Qwest had started using capacity swaps on its fiber-optic network to boost its revenues. Normally Qwest would sell or lease strands of fiber on its network to other companies and account the revenues for these 20-year deals over two decades. In an apparent attempt to cook the books, artificially inflate the revenues, and give an appearance of success and growth, Qwest decided to start booking the revenue and cash flow from the sale of fiber strands in one lump sum.

So, for instance, if the company had a deal for $100 million over 20 years, it would normally book $5 million in revenue every year. Under the new accounting, Qwest would simply book the whole $100 million as revenue, even though it would not get paid in full for about 20 years. Some managers at Qwest tried to put an end to this practice, but they were swept aside by the then chief financial officer Robert Woodruff. These managers approached Qwest's audit committee but were essentially ignored.[33]

Then, in March 2001, Woodruff resigned, perhaps worried about his own skin. He left with $29 million that he made from selling his shares and options. In came Robin Szeliga as the new chief financial officer of Qwest. Even she started having qualms about the whole mess. In an August 2001 memo, she wrote that officials had to be more careful with capacity swaps and not play loose with the rules. She had not even hit the send button on her e-mail message when she discovered documentation showing a side deal, another capacity swap between Global Crossing and Qwest Communications. It came from the computer of Afshin Mohebbi, then Qwest's chief operating officer.[34]

There was a lot of internal back-and-forth on the issue of accounting.

Meanwhile, Qwest stock had started going south. On August 31, 2001, Qwest was trading at $21.50 a share and was still under selling pressure. By the end of 2001, Qwest was trading at $14.13 a share. The year 2002 was going to prove to be disastrous for Joe Nacchio. In January 2002, Global Crossing filed for bankruptcy, which opened a whole telecom can of worms. On February 12, the SEC subpoenaed Qwest for some documents, and a month later launched a full-scale investigation into the company.

In April 2002, an unnamed employee sent a secret memo to the board of directors, stating that top people at Qwest, including Nacchio, had "set goals and targets that are impossible to obtain without engaging in unethical or illegal acts."[35]

This brought more attention to the company, and on June 16, 2002, the board of directors knew it was time to cut Nacchio loose. Qwest asked Nacchio to leave. By then, founder Anschutz had sold nearly $1.5 billion of shares, and Nacchio had sold about $250 million. But don't cry for underlings like former executives Lew Wilks ($46.6 million), Steve Jacobsen ($44 million), and Robert Woodruff ($29 million).

"I know these are big numbers. I'm neither apologizing for it nor embarrassed by it. I should be allowed to make more than a second baseman. I create more economic value than they do," Nacchio said.[36] He was riding into the sunset with his pockets stuffed with cash. So what if former US West employees' pensions were decimated, thousands were going to be out of work, and more would wonder what happened!

3

ONCE A JUNKIE, ALWAYS A JUNKIE

Gary Winnick, the son of a Long Island caterer, craved success, wealth, and attention. He got all of them, especially the attention, when the company he founded, Global Crossing, became one of the more grisly crashes of the telecom bust. Global Crossing, which hoped to build cross-continental optical networks, would become a case study in the excesses of the telecom boom. In less than five years, the company went through five chief executives, played with $15 billion of investor money, took on debt of $8 billion, and built a largely unused network that spans 85,000 fiber miles. Welcome to bankruptcy at light speed.

Winnick learned about the broadband business from a videotape that showed how fiber was laid under the ocean.[1] Even though the tape wasn't called "How to Be a Billionaire in 12 Months," it might as well have been, for Winnick went on to become a broadband billionaire, faster than anyone in U.S. history. He has since become a prime symbol of the corporate greed that overran the United States in the late 1990s.

■ ■ ■

Gary Winnick was born in Roslyn, New York, a small town on Long Island. His father, Arnold Winnick, was in the food and catering business and suffered a terrible blow when his business went bankrupt in 1960. He died when Gary was 18. The young Winnick adored his father. In later years, he would hire a Los Angeles–based film company, VDI Multimedia, to digitally remaster a tape of clips from his father's life, and give out the tape as a Christmas present to his close friends and family.

Gary graduated in 1969 from C.W. Post University in Long Island (now known as C.W. Post Campus of Long Island University) with a bachelor's degree in Economics. After college, he had a series of jobs, including one as a furniture salesman, before he joined the infamous junk-bond shop Drexel Burnham Lambert in 1972 as a bond salesman. He caught Drexel's junk bond powerhouse Michael Milken's eye in 1978 when Milken was moving his operations to Hollywood. And when Milken asked him to go to California with him, it took Winnick less than a second to say yes, pack his bags, and get the missus on the first flight to Los Angeles. It turned out to be the ride of a lifetime. Winnick spent seven years learning the art of leverage from the kings of the business, Milken and Leon Black, and sold a lot of junk bonds to gullible savings and loan companies.

In 1985, his coffers sufficiently lined, Winnick decided the time was right for him to start the firm he had been thinking about, with Milken's blessings, of course![2] He set up Pacific Asset Holdings, an investment partnership, in offices that were right across from Milken's West Coast headquarters, at the glamorous address of 1901 Avenue of the Stars, in Los Angeles's Century City neighborhood.

Winnick put some $30 million of his gains from junk bond sales into the partnership, Drexel Burnham kicked in $40 million, and another $45 million came from the famously wealthy Bass family and some of their investment vehicles. As a reward for seven years of loyal service, his former boss Milken helped Winnick put together a mega-million-dollar fund.[3]

Winnick was all set to go forth and conquer the world. But his debut in the high-stakes world of leveraged buyouts was inauspicious. He made a failed attempt to buy Western Union in the late 1980s, and subsequently stayed out of the public eye for many years—in fact, right up until he hit upon the idea that would become Global Crossing. Winnick's career has been so massaged by public relations half-truths that no one (except Winnick himself) is certain how and when Winnick stumbled upon the idea of getting rich off broadband and optical networks.

But stumble he did, and the salesman in him knew that this was his chance for a makeover that would make even Ricki Lake proud. Not that he knew anything about telecom. Before founding what would become Global Crossing, Winnick's only connection with the telecommunications industry was tangential, at best: His Long Island neighborhood had been carved out of an estate owned by AT&T financier Clarence McKay.

Public relations hype and the half-truths perpetrated by Winnick's lackeys credit Winnick with hatching the idea of Global Crossing. What is less well known is that it was two AT&T executives, Bill Carter and Wallace "Wally" Dawson, who first crafted a business plan that proposed laying optic cable under the Atlantic Ocean to connect Europe with the United States.[4] With telephone traffic increasing, the duo was confident that capacity on this network could be sold quite easily.

Just before Christmas 1996, Winnick was busy planning a trip to Hawaii to celebrate his 24th wedding anniversary.[5] His partner, David Lee, at Pacific Capital Group, a private partnership and investment bank, was meanwhile kicking the tires at AT&T, trying to figure out if there was any business they could buy from Ma Bell, which then was in a state of upheaval.

AT&T's Carter and Dawson proposed that they would use a fiber-optic cable to connect the United States with the United Kingdom, if Winnick and company could raise the $750 million required for the task. Winnick had no idea what the duo were talking about, so he got hold of a videotape that showed how fiber was laid under the ocean. He used a similar videotape to impress investors later. "That's how I learned how it was done," he later cavalierly told *Forbes* magazine.[6] This naiveté should have been a signal to everyone that they were best off staying away from his company. To be fair, it was a good business opportunity, but the man selling it clearly didn't have a clue. But those were the days of unfettered optimism, untempered by caution.

When AT&T sold its submarine business—a division that used specialized ships to lay down fiber-optic cables on the sea bed—to Tyco International in April 1997, Winnick and Lee lured away its management team, which included Carter and Dawson, and started Global Crossing. The company, which was started with $15 million of Winnick's money, proposed that it would eventually lay fiber 3,660 meters beneath the choppy Atlantic and connect the United Kingdom to North America. This cable, no thicker than a garden hose, would run from Brookhaven, New York, to Land's End, England, and ultimately to Sylt, Germany, and would be able to carry thousands of phone calls simultaneously.

Winnick's business plan claimed that a transatlantic network would eliminate a communications bottleneck and allow the company to sell that bandwidth to long distance companies. These companies would, in turn, greatly reduce the price of cross-Atlantic telephone calls. Soaring demand coupled with a shortage of capacity created a perfect opportunity for anyone to make a killing—provided they had enough cash to get such a capital-intensive project off the ground.

With the telecom industry being deregulated, it was hard for Winnick not to lick his chops at the thought of how much capacity he could sell to all those established companies and the new independent telephone carriers that were cropping up like mushrooms after a monsoon. The project would be tremendously expensive, but Winnick justified the price by saying that people would pay to remove the serious bottlenecks that existed on transatlantic cables. He took his plan to his Wall Street pals from his Drexel days. In late 1996, Winnick got in touch with Bruce Raben, then a managing director at Canadian Imperial Bank of Commerce (CIBC). Raben, who had worked with Winnick at Drexel, took the proposal back to his colleagues in New York.[7]

In March 1997, CIBC invested $41 million in exchange for 25 percent of Global Crossing. It was an insane valuation, for there were no real assets in the ground—or under the sea, for that matter. It was the best bet the Canadian bank ever made. By the end of 2001, CIBC had turned that investment into $1.3 billion—a return of over 3,000 percent.[8] The bank also led the syndicate that loaned Global Crossing $482 million in 1997 and later helped out with another $850 million in debt. In exchange, CIBC got five seats on Global Crossing's board. But there is more than one way to skin a cat. According to the research group Thomson Financial, CIBC also raked in $20 million for underwriting the Global Crossing initial public offering and another $12 million in consulting fees.[9]

In the early days of Global Crossing, its business plan made absolute sense. It was later that the troubles would begin. There truly was a bottleneck under the sea, and transatlantic phone calls were prohibitively expensive. Using Global Crossing's pipe, the costs could go down drastically—to about 38 cents an hour. And that meant phone companies like AT&T and MCI could offer 5-cents-a-minute phone connections to London. Cheaper calls meant more people would talk more, and that in turn would fill up the network, thus forcing phone companies to buy more capacity from Global Crossing. Within three months of launching Global Crossing, Winnick's coterie had raised enough money, and within 14 months the transatlantic cable was being laid down.

Now Available: The Brooklyn Bridge!

While the cable was being laid, Winnick took his message to a group of phone company executives who gathered in October 1997 at the Marriott Hotel on Manhattan's East Side.[10] At the meeting, Winnick, the consummate salesman, told the executives that he was willing to sell his connections for $8 million a circuit, versus the prevailing price for $20 million a circuit. While he did not lose any money on these deals, he left a lot of money on the table. Given the scarcity of bandwidth, the phone company executives would have gladly paid double Winnick's asking price. "We even offered an early-bird special that was less than that," said Harold Grossnickle, a former Global Crossing executive.[11] If anything, this departure from basic rules of supply and demand should have raised a red flag among investors. Even a sophomore majoring in philosophy could see that discounting prices when capacity was scarce was not a sound economic decision. But phone company executives were certainly not going to complain, for they were getting a bargain.

What ensued was a feeding frenzy. In a couple of hours after the meeting, Winnick had sold enough capacity to recoup half of his $750 million bill. This was easier than selling Botox to Upper East Siders. Still, all wasn't lost, if Winnick had stopped there. But he didn't. At the Marriott meeting, he is said to have hand-drawn a map of the world and crisscrossed it with fibers, diagramming a network that would reach the farthest corners of the world. Unfortunately, that crude drawing became the company's global business plan. Its grandeur was exceeded only by the time required to make it profitable.

Global Crossing's ambitions knew no rational bounds. By the end of 1997, another connection between New York and the Virgin Islands was announced. It was followed by a torrent of press releases announcing cables along North America's western coast down to Panama and across the Pacific. Global Crossing would eventually own more than 85,000 miles of fiber.

"When we began to construct our 100,000-mile fiber-optic network, it seemed as though there was simply not enough fiber-optic capacity to satiate the appetite of a world that would become committed to transmission with ever increasing and enormous amounts of voice, data, and video traffic," Winnick later said. "Our vision was one of innovation and competition, to be the first company out of the gate in building a global network to meet demand and to provide the best possible service to our customers."[12]

But the whole idea wasn't even original. It seemed Winnick had simply taken over the business plan of a Global Crossing rival called Project Oxygen.

Gasping for Air

In 1997, Neil Tagare organized what became a seminal event in the history of broadband. On December 7 that year, more than 500 representatives from 330 phone companies spanning 150 countries descended on Las Vegas to attend a telecom version of *The Predators' Ball.* Everyone was there—American companies like GTE and AT&T, plus representatives of telecom companies in Israel, India, Egypt, and Chile. For four days, telecom executives pondered Tagare's vision, which was quite simple: Connect all of the world's populated areas with a vast fiber network, and then sell bandwidth on an as-needed basis. He called this Project Oxygen.

Project Oxygen's fiber-optic cable network would span 198,400 miles, would be able to carry at least 320 gigabits of data per second, and would link every continent except Antarctica. The project would cost a whopping $14 billion to construct, and would touch 265 end points in 171 countries. "We call it Project Oxygen because carriers who join us will survive, and carriers who don't join us won't survive," Tagare boasted.[13] At that time, carriers bought capacity on international networks in 25-year contracts. And here was a 32-year-old former consultant proposing to upend the whole industry order.

The four-day event cost $2 million, and attendees were given a chance to drive in a NASCAR race car, get racing lessons, and even dance with the cancan dancers. It is rumored that it was one of the best weeks for Las Vegas' ladies of the evening. Over those four days, Tagare managed to entertain, cajole, and bully carriers into buying $3 billion worth of capacity in a network that did not even have a single fiber under the sea. "We presold $3 billion of capacity but I wanted to do $7 billion. I could not close it, but a lot of people bought my story," said Tagare. The reasons were simple—most of the world's submarine cables were owned by a handful of companies, including AT&T, MCI, British Telecom, and NTT of Japan, and carried $90 billion worth of phone calls every year. Tagare wanted to let other phone companies in on the action. This model would effectively bust the trusts. "I will offer the same quality of service to the telephone company in Brunei as AT&T does, for a lot less money," Tagare said.[14]

Telecom industry observers were skeptical about the project's broad scope. "The incumbent global communications players may well pull out all the stops to deflate Project Oxygen," Graham Finnie, an analyst with the consultancy The Yankee Group, told *Red Herring* magazine in 1997.[15] "Because, if this works, the cost of trivial amounts of bandwidth, or international phone calls, will fall so low that it won't be worth billing end users, and a $90 billion revenue stream will dry up in five to seven years' time." Still, Tagare became the toast of the town. *Fortune* magazine had named Project Oxygen and its primary promoter, Tagare's CTR Group, as one of the "cool companies of 1998." CTR stood for Concept to Reality!

From Bombay to Broadband

Sunil "Neil" Tagare, a native of Bombay, India, had emigrated to North America to pursue a master's degree in engineering. Upon graduation, he joined Kessler Marketing Intelligence, a telecom consultancy based in Newport, Rhode Island. There, in 1989, he hit upon the idea to build a fiber link that connected various ports in Europe and Asia. That became FLAG (the Fiber-optic Link Around the Globe) Telecom. The idea behind that venture was to build a network that could carry voice traffic emerging from countries less served by the mainstream telecom infrastructure. NYNEX, the northeastern Baby Bell, became the initial backer of the project. FLAG ran 25,000 kilometers of cable from England to

Japan—a geographically challenging task—and was profitable very early on. That project helped bolster Tagare's reputation in the industry circles. No wonder he nurtured such dreams for Project Oxygen, an audacious project so huge, that it would cover the entire planet in a sea of bandwidth. By the time Tagare got off the ground, Winnick's Wall Street connections had brought Global Crossing enough money to build the first leg of the global network.

Tagare's backers included Alcatel Submarine Networks and Tyco Submarine Systems, Ltd., as well as Japan's NEC Corporation, NTT International Corporation, Sumitomo Corporation, and Mitsui & Company. Lucent had promised $1 billion in equipment. J. P. Morgan promised to raise a whopping $3 billion in private equity. Of course, it was all based only on a PowerPoint presentation at the time—one of many start-ups that raised billions of dollars on the shallow merits of a slide show. "A carrier can buy as much capacity as it needs today and then scale up and down," Tagare told tele.com, an online publication tracking the telecom industry.

Project Oxygen, which was much bigger than FLAG in its scope, ran into problems. Tagare refused to give up any equity, and by October 1999, Project Oxygen had barely managed to raise $250 million, most of it coming from two Israeli companies—Bezeq International, a telecom company, and Elbit Medical Imaging—in exchange for a small portion of Project Oxygen.

But despite the absence of any network, Tagare's efforts had not gone unnoticed. Gary Winnick, who had originally imagined that his company would connect the United States with Europe using a transatlantic cable, decided that it was time to take a page out of Project Oxygen's business plan and expand his scope and ambition. He slowly started to use his influence with the equipment makers, contractors, and Wall Street bankers to cut off the air supply to Project Oxygen. Winnick reportedly even went so far as to persuade TyCom, a construction company that was to build Project Oxygen's cable links, to pull out of the project. TyCom had been the undersea construction business unit of AT&T, which Ma Bell had sold to the now scandal-ridden Tyco International.

Tagare was fighting a losing battle but refused to give up. He managed to convince AT&T to take a 10 percent stake in the company, and even was rumored to have lined up other investors such as Telstra, an Australian phone company. But his main source of funding was Pacific Century Cyber Works (PCCW), a Hong Kong–based Internet incubator and investment vehicle for Richard Li, scion of one of Asia's richest families, with varied interests including satellite communications systems.

Unfortunately, Li's commitment to Project Oxygen was short-lived. In a stunning development, in February 2000, PCCW bought Cable & Wireless's Hong Kong Telecom business for a whopping $38 billion. Cable & Wireless already owned an undersea network, and PCCW did not need Project Oxygen anymore. On April 9, 2000, Li called Tagare, who was asleep in his New York hotel room, and terminated his commitment.

As late as the summer of 2002, Neil Tagare was still bitter about the whole experience, blaming Winnick for the eventual death of Project Oxygen. "Winnick could play dirty—he called the suppliers who had funded me and told them that he was going to withdraw them from the supplier lists. He is an extremely ruthless guy," says Tagare, who was no match for the wily Winnick. Project Oxygen never really got off the ground—it just remained a dream for Tagare.

Buy Something, for God's Sake

With Project Oxygen out of the way, Global Crossing was unchallenged and started to build a massive network. From 1996 to 1998, the demand for bandwidth had grown at breakneck speed. The growing number of dot-coms wanted fast connections to their computers, and so did the Jane and John Does who wanted to get on the Net. Other countries were also getting into the act. In August 1998, Global Crossing sold 21 million shares to the public at $19 a share, raising a whopping $399 million. The shares would soon change hands at $64.25 a share, and the company would ultimately be worth around $49 billion in its heyday, which was more than General Motors and other long-established American blue chip companies.

Global Crossing's competitors, like Qwest and Level 3 Communications, were concocting their own international plans. Even as early as 1999, a huge capacity glut seemed to be looming. Global Crossing executives sagely figured out that the game was going to be over soon, but they needed time to cash out. So they started announcing plans that made little sense in terms of revitalizing or rescuing the core business but did a lot to inflate the stock price. One such plan was to start an Asian operation called Asia Global Crossing, which would provide bandwidth to the booming Asian economies in the Pacific Rim and Southeast Asia. Asian investors like Softbank signed up as initial backers, and Global Crossing's stock kept rising, because investors naturally saw expansion as a sign of good health.

Despite the looming capacity glut, Salomon Smith Barney analyst Jack Grubman, who fancied himself something of a telecom soothsayer, urged Winnick to go on a buy-and-grow path, a mantra he had preached to other telecom chief executives such as Joe Nacchio of Qwest and Bernie Ebbers of WorldCom. Winnick took his advice and decided to get into the local market. In 1999, he made a $36.5 billion bid for US West, and another bid for Frontier Communications of Rochester, New York. He did not get US West, but $8.1 billion got him Frontier. (Global Crossing would later sell its local business to Citizens Communications of Stamford, Connecticut, for $3.7 billion in 2000.) Salomon's fees for advising Winnick: $16 million.

Emboldened, Winnick now wanted to do a massive global buildout, raising billions in one shot that would complete his global network. On October 29, 1999, the company raised a whopping $3 billion—$1 billion in stock and another $2 billion in debt, all thanks to J. P. Morgan Chase's head of investment banking, James "Jimmy" Lee. In the investment banking circles, Lee was known as Jimmy Fee because he would undercut rivals on investment banking fees. James's bosses must have been pretty pleased with him for wrestling away business from Salomon, which was owned by the rival bank Citigroup. Chase's share of the telecom merger advisory business soared from the low single digits in 1998 to 17 percent in 2000, thanks in large part to business thrown the bank's way by Global Crossing, *Fortune* reported.[16]

Flush with cash, on November 18, 1999, Global Crossing announced it would do a 50-50 joint venture with Hutchison Whampoa called Hutchison Global Crossing. And while the going was good, the company decided to raise more money. Global Crossing sold another $630 million worth of convertible stock.

With its revenues topping $1 billion, the company announced in February 2000 that it was entering a deal with Level 3 Communications. Level 3 would buy capacity on Global's transatlantic cable, and Global Crossing would take 50 percent of the cable capacity being built by Level 3. A couple of days later, the company snapped up an enterprise network services provider, IPC/IXnet, for around $3.6 billion in stock. Global Crossing wanted to get into the enterprise business, and IXnet, whose customers included top names on Wall Street, could deliver such customers.

The Frontier and IPC/IXnet acquisitions were just a means to fill up Global Crossing's empty pipes. Competition was increasing from Level 3, Qwest, and a dozen other companies that had jumped into the global network carrier business. Even in early 2000, the bandwidth glut was having a negative effect on the operations of most broadband operators. All were aggressively competing on price, their only leverage. And they were not as easy to get rid of as Neil Tagare's Project Oxygen. "In the early years, demand for global broadband conductivity was insatiable. Global Crossing's success attracted many competitors with their own financial backers, eager to replicate Global Crossing's reach," said Patrick Joggerst, former vice president/president of carrier services.[17]

Money for Nothing

Global Crossing obviously got a big boost from Wall Street and Jack Grubman, who boosted the stock no end. Always ready to please, Grubman was said to be in Winnick's back pocket. Global Crossing insiders needed his boosterism—according to *Fortune* magazine, insiders disposed of about $4.5 billion in stock. This included $36 million made by Global Crossing's gray-haired sage and co-chairman Lodwrick Cook. The former chairman of Atlantic Richfield sold $36 million of stock, while other members of Winnick's inner circle, Barry Porter and David Lee, sold $516 million worth of shares, the magazine estimated.

After the Global Crossing IPO, Winnick's stake was worth more than $6 billion. It took John Rockefeller 25 years to make his first billion; Winnick made six times as much money in two years. It took Bill Gates 15 years to make that much money! But, unlike Rockefeller or Gates, Winnick made his fortune from a house of cards. He began to sell Global Crossing shares from the get-go. His final take-home was $735 million, or roughly 50 times the initial $15 million he plunked down to start the company. At one point Winnick is said to have boasted that he was the richest man in Los Angeles.

Some of the biggest moneymakers in the Global Crossing bonanza were the five men who warmed the chief executive's chair over a period of 36 months. Winnick apparently did not want to deal with day to day affairs, and was happy being the chairman of the board. Jack M. Scanlon, a former Motorola executive and erstwhile vice chairman of Global Crossing, was the chief executive for all of ten months, but he still walked away with over $170 million. Robert Annunziata, Leo Hindery, Thomas Casey, and John Legere followed Scanlon in the executive suite and all had multimillion-dollar paydays. The most memorable line describing the phenomenon came from the *Denver Post* business editor, Al Lewis, who wrote, "One way to get rich quick in America is to become CEO of Global Crossing." One wag likened Global Crossing to the New York Yankees, where owner George Steinbrenner's constant meddling resulted in the well-known revolving door for coaches.[18] A former Global Crossing executive told *Fortune* magazine: "Gary was running the place like he was still at the bond desk at Drexel in the '80s. People were bought, and money was thrown around."[19]

Take the case of Thomas "Tom" Casey, a former investment banker who fell in with the Global Crossing crowd in 1999. He was bought for $20 million. Winnick could not resist boasting about it, and told the news program *News Hour with Jim Lehrer* about it. "I said, 'Tom, what does it take?' He says, 'Gary I'm going to throw out a number, it'll never happen.' 'Give me a number, Tom.' '$20 million.' I say, 'When can you start? I'll write the check,'" boasted Winnick.[20]

Casey and his wife were close friends of the Winnicks. The two couples vacationed together in Italy. But Casey fell out of favor when he could not fix the problems at Global Crossing. He was shoved aside, and in came John Legere. In recent hearings, Winnick all but blamed Casey for Global Crossing's collapse, who has not spoken with Winnick since he was fired in October 2001.[21] The tussles with Hindery and Casey are indicative of Winnick's inherent insecurities.

Winnick's insecurities aside, the Global Crossing culture was all stock all the time. No one personified the obsession with the stock price better than Joseph P. Clayton, who was the vice chairman of Global Crossing and also served as president and CEO of Global Crossing's North American regions. Clayton used to drive around in a car with a license plate that said, "GBLX 100," which indicated Global's stock ticker and target price. Before assuming his new role at Global Crossing, Clayton was president and CEO of Frontier Communications, and his license plate then read "FRO 50"; today, he is CEO of Sirius Satellite Radio. Through a spokesperson, Clayton said that the license plates were for motivating employees.

Clayton owned about 2 million shares and options of Global Crossing, worth more than $50 million at the very least. Like Clayton, many fell under Gary's spell and were members of the Winnick Fan Club. Those who were not, like Hindery—who Winnick at one occasion called a "pathological liar"—were disposed of quickly.

Inside Gary's World

Those who worshiped Gary Winnick were rewarded for their loyalty. They would get to fly in private choppers to Watkins Glen, a NASCAR racing track in upstate New York, to see NASCAR races. The company would fly them to exotic locales in a fleet of private aircraft that included Lear Jets, helicopters, and a Boeing 727. Helicopters were used exclusively to fly executives from New York to Global Crossing's Madison, New Jersey, offices. Lodwrick Cook and Winnick would go golfing in Monterey, California, so regularly in the company jet, that mechanics dubbed it the "golf-stream."

The company was headquartered in Hamilton, Bermuda—for tax reasons, of course—but the company's executives worked out of Los Angeles and satellite offices like the one in Madison, New Jersey. Winnick's gang hung out at the former Music Corporation of America building, after giving it a makeover and renaming it—what else—Global Crossing Plaza. The funding for this came from Global Crossing, the Pacific Capital Group, and Colony Capital (another private investment firm with backing from Winnick). The office was on the second floor of the three-story MCA building. There were swarms of tall, beautiful blondes in the office. No one seems to know what their role was, except to run out for shoe-shopping runs at nearby Rodeo Drive and for two-hour lunches every day.

The message coming out of the executive chambers was that anything goes and money is no object. People were spending money like drunken sailors, according to insiders. Visiting executives were put up in the nearby Beverly Wilshire Hotel—the hotel where Julia Roberts ran amok in *Pretty Woman*—at $2,000 a night. No one was shy about flaunting their wealth. Winnick gave away trinkets like Aston Martins and Rolls Royces. He was treating the company like his personal piggy bank.[22]

Winnick, who was clueless about the telecom business at the beginning, now thought of himself as a network guru. According to some former executives, Gary one day strode into the office and demanded that an OC-3 connection be installed to his house. OC-3 is a technical term for a fiber connection that can transmit data at 155 megabits per second; it's roughly equivalent to 100 times the speed of the T-1 connection that typically connects corporations to the Internet. The story, according to company insiders, is that they hooked Gary up with a T-1 connection, and he never knew the difference.

Winnick wanted to project the image of power through his office. He often played up his blue-collar roots, just as a reminder to others of the magnitude of his success. While it has been widely reported that Winnick's office was a replica of the Oval Office, in reality it was a rectangular office with a slightly curved alcove at the back. Winnick would stay in the office late at night, pacing up and down and looking at the expensive, dark wood paneling. He would almost always be on the phone. Expensive art on the walls, a sitting area for guests, and a big empty table were the only furnishings in his office. A couple of flat-screen televisions, muted, would beam financial news all day long. Gary would dress mostly in custom-made suits with a Global Crossing pin on the lapels—Hickey Freeman when formal, and Arnold Palmer when casual.

"(His) hair was perfect, and he was always perfectly groomed and very charming," recalls one former executive. Winnick was a chubby man who was always struggling with his weight. He used to run staff meetings while huffing and puffing on a treadmill he kept in his office. When things got bad, he would go on an eating binge and get bigger and bigger. The more pressure on him, the fatter he got.

House of Cards

Even when Global Crossing began to run into trouble, Winnick's personal spending continued to run amok. In 2000, he bought a 15-bedroom 23,000-square-foot mansion in Los Angeles' tony Bel-Air neighborhood for a whopping $65 million—the highest price ever paid for a single-family home in the United States. And he was going to spend $30 million on renovations! This was in addition to Gary's Malibu home, which was known as Chez Malibu, where he lived down the beach from Barbra Streisand. "Money's no fun unless you spread it around," Winnick told *BusinessWeek*.[23] This was the property which would eventually become Winnick's house of pain and a symbol of his greed.

This legendary Bel-Air mansion, the Casa Encantada—enchanted house—is located on Bellagio Road, right next to the Bel-Air Country Club. It has a stunning view of Los Angeles. The *New York Times* described it as being like a house on a hill in the center of Central Park, with a view of Manhattan. The house's first owner was equally infamous. Hilda Bolt, a widowed Los Angeles socialite wannabe, had purchased the property for $100,000 in 1934. A nurse from New York City, she had married one of her wealthy patients, who didn't last long. Bolt spent $2 million, or about $24 million in today's money, on the house that she wanted to share with her chauffeur and second husband, Otto Webber. She hired T. H. Tobsjohn-Gibbings to do the interiors. He was the "it" designer of the day, having created interiors for Doris Duke and Elizabeth Arden. She did not get to enjoy the house, as bad investments and gambling debts forced her to sell it for $225,000, to Conrad Hilton, the founder of Hilton Hotels, in 1948. Hilton stayed there for 19 years, and eventually sold the house to David Murdock, the chief executive of Dole Food Company.

Murdock entertained presidents Richard Nixon, Ronald Reagan, and George Bush (Sr.) in the Casa Encantada. It was while attending one of these political functions and fund-raisers that Winnick fell in love with the house and made many desperate bids to buy it, raising the bid each time, until Murdock couldn't say no anymore. Winnick bought the 64-year-old estate from him, and entrusted well-known designer-decorator Peter Marino to do a makeover on the 43-room house, which includes 17 bathrooms, tennis courts, and a swimming pool. It is said that the house has a kitchen with six sinks, while the master bedroom has separate massage, sitting, and shower rooms attached to it. The house, which sits on an 8.4-acre spread, is hidden from the outside world by trees, and the grounds are big enough to fit three baseball fields. When Global Crossing's star and stock were falling, Winnick was undertaking a multimillion-dollar renovation.

"The company's going down the tubes, and he's flaunting his money and spending millions of dollars on the house," lamented Michael Nighan, a former Global Crossing employee, to the *Los Angeles Times*.[24]

Buy Some Respect

But Winnick didn't spend all of his money on himself. As his wealth grew, so did his philanthropy. Overseeing Winnick's philanthropy—and helping his social-climbing—was Rosalie Zalis, executive director of the Winnick Family Foundation. Zalis, a powerful person in Los Angeles, was a former publicist for California Governor Gray Davis and was sometimes described as Gary's other mother. She introduced Winnick to local Democrats, including Governor Davis. She connected him to local charities and also with the Simon Wiesenthal Center, a Jewish human rights organization whose Jerusalem branch would eventually be named after Winnick. He gave millions to various organizations that included everything from his alma mater, C.W. Post University on Long Island, to the Los Angeles Zoo.

In addition to Zalis, Winnick hired Lodwrick Cook to co-chair Global Crossing, largely because Cook was connected to powerful people, including the Bush family and other Washington power brokers. "To him it was all about image, and he would consult Zalis about everything," recalled a former Global Crossing executive.

This kind of philanthropy gave Winnick an entrée into high society. He played golf with then President Bill Clinton and received calls from Buckingham Palace. California Governor Gray Davis called him a personal friend. In what now seems a sickening homage to the façade Winnick had built from the ground up, Governor Davis credited Winnick and Global Crossing with fueling California's economic expansion. On November 28, 2000, at the opening of the Global Crossing Plaza in Los Angeles, Davis said, "Global Crossing, the Pacific Capital Group, and Colony Capital have helped to fuel the greatest economic expansion we have ever known. Now this magnificent building will serve as a breathtaking visual reminder of their many contributions to California's economic landscape."[25]

Others politicians benefited from the Global Crossing association as well. Former President George Bush (Sr.) and Terry McAuliffe, a Clinton crony and chairman of the Democratic National Committee, got rich off Global Crossing. Bush Sr. made his money the old-fashioned way—he gave speeches at various Global Crossing events and, in lieu of $80,000 in cash payment, opted for Global Crossing shares that were worth about $14 million at their peak. At a White House event, President George W. Bush referred to Winnick as the man who gave his father the stock.

McAuliffe got $100,000 in shares for making an introduction to President Clinton. McAuliffe's stake during Global's heyday was worth $18 million. Unlike Bush Sr., McAuliffe bragged about the million she had made from Global Crossing in an interview with *New York Times* reporter Jeff Garth. Winnick was definitely an equal opportunity corruptor, bipartisan in his largesse. Political patronage was part of Global Crossing culture. Having learned his lessons in the 1980s, Winnick knew that in order to escape political scrutiny he needed to buy political protection. It cost him and his company less than $5 million, a cheap price to pay for politicos.

In 1999, when Global Crossing wanted to build a trans-Pacific fiber network and needed approval from the Federal Communications Commission, the company hired Anne Bingaman, who had been head of the Justice Department's antitrust division from 1993 to 1996. For $2.5 million, Bingaman successfully lobbied on behalf of the broadband company. The fleecer-in-chief of Global Crossing, his cronies, and some of the other companies he controlled gave President Bill Clinton and the Democrats at least $1.2 million.[26] Republicans received a little bit more, including $185,000 personally from Winnick.[27]

In August 2000, Winnick paid for a luncheon at the National Democratic Institute, an event attended by President Clinton, Secretary of State Madeline Albright, and a host of other Washington insiders.[28] When Madame Albright got up to speak, Winnick, moving surprisingly fast for a man fighting a losing battle with his waistline, placed a baseball cap featuring the corporate logo of Global Crossing on her head. The consummate salesman knew that press photographs would show a respectable member of the Clinton administration wearing his company's logo, and the message would go out: We are connected, we have protection, and we can pretty much do anything.[29] It surely would impress the Asians and Europeans who place a lot on political patronage. Winnick, after all, was building a global company.

Albright was not the only one who was trotted out to prop up the already crumbling edifice of Global Crossing. In 1998, the firm hired the lobbying services of Norman Brownstein, one of the nation's most influential attorneys and an adviser to political notables in both major parties. Brownstein was a Global Crossing board member until November 2001. In April 2001, when Global Crossing was about to go down in flames, former Defense Secretary William Cohen joined the board of directors of Global Crossing and its subsidiary Asia Global Crossing, four months after leaving the Pentagon.

Among other political recipients of Winnick's largesse were several key figures involved with telecom regulation. John McCain, the Republican Senator from Arizona and eventually a presidential candidate, got $31,000, while Representative Edward J. Markey, a Democrat from Massachusetts, settled for $12,500. McCain, on behalf of Global Crossing, presumably wrote to the Federal Communications Commission to encourage the development of undersea telecom cables. In 2000, the company spent nearly a half-million dollars on each party's convention. It even provided Web hosting and Internet connections for the Republican convention—free of charge, of course.

It was easy to overlook all this, given that business and politics are so closely entwined all over the world, whether in the United States, where such monies are called contributions, or in the third world, where they are more pejoratively referred to as *baksheesh* or graft.

The media, in those heady days of the telecom and technology bubbles, overlooked this backdoor deal-making. Charmed by his self-promotional spiel, magazines including the usually skeptical *Forbes* made Winnick their cover boy. In April 1999, *Forbes* magazine gushed about Winnick and what a maverick he was. The story, "The $20 Billion Crumb—Getting Rich at the Speed of Light," even talked about how Global Crossing was profitable in its first year—if you did not take into account the $139 million Winnick and others took for "consulting" services provided to the company.[30] What consulting service could he be providing, when this was the same man who only a few short years ago may have assumed that fiber meant Metamucil? More importantly, why was Winnick earning consulting fees if he was the chairman of Global Crossing? Ironically, the man who initially learned about broadband from a simplistic videotape[31] was now successfully passing himself off as the premier soothsayer of a networked future.

And Now, Global Double Crossing

The future of Gary Winnick and Global Crossing was decided the day they lost the battle for the mountain Bell, US West, in the summer of 1999. Winnick satisfied himself with a consolation price by acquiring Frontier Communications, a Rochester, New York–based company, for $8.1 billion. Now, Frontier was not a bad company to own. It had been around for about 100 years and was known as Rochester Telephone until 1995. It had lots of fiber in the United States, and that could provide links to many different cities across the country. For Global Crossing, which was mostly international in its scope, it was a nice way to expand service to North America. However, by the time the merger closed, the broadband market just tanked, and so did Global's fortunes.

Even inside the company there was a culture clash, and so much confusion that sales and marketing people from different groups were tripping over each other. In May 2000, the geniuses at Global Crossing decided to split Frontier into two—they kept Frontier's national backbone network, but sold the local phone business to Citizen Communications. They should have done the reverse, but then Winnick and his crew never really knew the business. Nevertheless, the deal brought Global Crossing about $3.65 billion in cash. Not bad—but most of that money was used to pay down Global Crossing's debts, according to documents filed with the SEC. The company also recorded a loss of $208 million on the sale.

The free-spending ways of Winnick and his coterie, along with a downturn in the bandwidth market, were making life hell for Global Crossing. A bubble of demand was created by thousands of dot-coms, which were bandwidth hungry and were expanding like crazy between 1996 and 2000. However, when the stock market bubble burst in March 2000, the dot-com dominoes started falling. Overcapacity in the bandwidth market, combined with a sharp decline in demand for bandwidth, was too much for broadband companies that had sprung up in late 1990s all around the world.

The "bandwidth barons," as the executives of these carriers were known then, figured out a new way to keep the hype alive—sell bandwidth to one another. It was the telecom version of the good old Ponzi scheme. Global would sell a billion dollars worth of connectivity to Company X, and Company X would turn around and sell an equal amount of bandwidth to Global Crossing, albeit somewhere else on their network. No cash ever changed hands. These swaps are different from the so-called Indefeasible Rights of Use. IRUs are legitimate and are a common practice in the industry, and are usually for long periods of time, almost 20 years. The companies get paid over that period of time as well. Swaps are simple exchange of capacity. But Global Crossing blurred the lines between swaps and IRUs.

But Global Crossing, desperate for revenue, decided to book IRU sales as immediate revenues and further compounded the malfeasance by booking purchases as "expenses" that would be spread over many years. This created a perception that its revenues were increasing, even though not a dollar was changing hands. But all that time, in 2000 and 2001, with bandwidth prices in a free fall, Global Crossing was running on fumes.

It found willing partners in other players because they all were having the same problem: no revenue, no customers, and very little demand. In the fall of 2000, Enron, Reliant, and Global Crossing had a capacity ménage-a-trois.[32] Global Crossing and Enron swapped capacity and services worth $17 million, and Reliant brokered the deal. Global Crossing booked revenue from capacity it sold to Enron, and Enron did the same, but using Reliant as a conduit.[33]

Global's other partner in slime was Qwest and Joe Nacchio. Qwest reported revenues of more than $1 billion from network capacity sales in 2001, but as it turned out, more than two-thirds of those sales were swaps, in which Qwest simultaneously purchased similar amounts of capacity from the purchasers of Qwest bandwidth. Global Crossing reported $720 million in cash revenues from the sale portion of these capacity swaps in the first and second quarters of 2001 alone.[34] The one-time foes were now indulging in a game of "you scratch my back, and I'll scratch yours."

Even as all this was happening, Winnick's crony Jack Grubman of Salomon, who also attended some board meetings of Global Crossing, was trying his very best to prop up the stock, which had sunk from a high of $61.38 in 1999 to $16 in November 2000 and to almost $12 by April 2001.

Turns out not everyone was comfortable with what was going on. At a meeting in April 2001, Tom Casey, then the CEO of the company, voiced his concerns. He pointed out that revenues could be down by a billion dollars. Others were predicting half-a-billion dollar shortfall in revenue for that quarter. In the same month, Grubman released a report titled "Don't Panic: Emerging Telecom Model Is Still Valid," and pumped up Global Crossing. In May 2001, he again said much of the same—"buy Global Crossing." As luck would have it, Winnick sold $123 million in stock that month. Put two and two together—and you get a sense of collusion. "The suggestion I sold stock based on information not readily available is not correct," a petulant Winnick later said.[35]

Whatever his misgivings, Casey, in response to an analyst's question on the earnings conference call, claimed that there were no swaps in the quarter. Roy Olofson, the company's former vice president of finance, who had joined Global Crossing in 1998 as the 40th employee, was incredulous at this blatant lying. "While I was on leave [from January 2001 to May 2001], I learned that Global was having a very difficult time meeting its first quarter revenue projections," said Olofson, who was the first one to blow the whistle on the scam. "I also learned that Global ultimately was able to meet its numbers in part due to some large, last-minute transactions where Global swapped IRU capacity with other carriers." Even though Winnick claimed he never really had anything to do with it, the *Wall Street Journal* has reported that in June 2001, Winnick discussed a $900 million telecom capacity swap with Enron's former chief executive, Jeffrey Skilling.

According to the *Los Angeles Times*, Winnick was aware of the swaps. A June 11, 2001, e-mail, indicated he had a conversation with Skilling about a deal that was never completed.[36]

Winnick may have pretended ignorance about day-to-day affairs, but the *LA Times* reported that he was involved intimately in the running of the company. He handled renovation of the Beverly Hills offices, including deciding on the color and light fixtures. Winnick made changes "to the cafeteria menu, ordered flowers replaced, chose the color of the company's logo and the font to be used on its letterhead and other documents," the *LA Times* reported.[37]

"I began to learn that there was a general sense of uneasiness about these swap transactions and in particular about a transaction with 360networks. Through discussions with various people, I learned that 360networks and Global Crossing had entered into a last-minute transaction wherein Global booked $150 million in cash revenues even though it had not received a penny in cash," Olofson added. The transaction originally called for Global Crossing to pay $200 million to

360networks, another long-haul carrier, and then for 360networks to pay Global Crossing $150 million. In reality, only $50 million would change hands, but even that did not happen because at the time there were rumors that 360networks was about to go bankrupt.

According to the *Wall Street Journal*, in March 2001, 14 executives from Global Crossing and 360networks met at the offices of Simpson, Thacher & Bartlett, one of Global Crossings's law firms. This was the second time the two companies were getting involved in swaps. Earlier the two companies had conducted a $180 million capacity swap. "There was a subtext of desperation," someone who attended the March 2001 meetings told the *Wall Street Journal*.[38]

Later, Olofson met Global Crossing CFO Joe Perrone—his boss—and discussed the financial condition of the company. "I took the opportunity to express my concerns about Tom Casey's statement in the quarterly conference call, that there had been no swaps in the first quarter, when in fact there appeared to have been a significant number and a substantial dollar amount of swap transactions. I also told him there were a number of people in the office concerned about the accounting for those swap transactions," Olofson later claimed.

Perrone apparently did not pay much heed, and on August 2, 2001, Casey once again publicly said that there had been no swaps in the quarter that ended June 30, 2001. Olofson sent a letter to Global Crossing's chief ethics officer, James Gorton, on August 6, outlining his concerns. Were company ethics a big concern, given that all the stinky stuff was coming out of the executive chambers? Gorton assured him that he would investigate the matter. "Perrone attempted to brush off my concerns. He stated that he had added some language to Global Crossing's press release regarding purchase commitments and that he interpreted the question from the analyst, to which Mr. Casey responded as referring only to transactions called Global Network Offers and not to capacity swaps," Olofson later told a House committee investigating Global Crossing. "At the time, I believed the company would investigate my concerns in good faith. I was wrong. Instead, they fired me."

But Olofson would be proved right, and soon the financial problems would become a full-scale crisis. The company attempted to get its act together but did not succeed. Global Crossing said that it would close

100 of its 600 offices, mostly in markets with multiple offices or where two offices perform similar functions, hoping that it could save the company about $160 to $170 million a year in operating expenses. (This begs the question, why were there two or more offices in multiple locations anyway?) By October 2001, the press had gotten a whiff of the scam and began asking tough questions. First in line was Elizabeth Douglass, a reporter with the *Los Angeles Times*. "Winnick said Global Crossing's results in the July-September period took a hit because the company turned away as much as $900 million in capacity 'swapping' deals with other carriers, labeling the proposed contracts 'bad business,' " Douglass wrote.[39]

At the time, the firm had just hired a new CEO, John Legere, who came to Los Angeles from Asia Global Crossing. The firm was bordering on insolvency, but Legere got an annual salary of $1.1 million and a $3.5 million sign-on bonus. In addition, he was going to get 5 million options and an annual bonus that would amount to 125 percent of his base salary. Asia Global Crossing forgave the $10 million balance of a $15 million interest-free loan Legere had received when he left his previous job at Dell Computer. Asia Global Crossing would also pay him an additional $2.75 million in severance. Even if one assumes that those 5 million options were by then totally worthless, it was a $24 million sign-on package. Not to mention, what on earth was Asia Global Crossing doing, giving someone a severance package when they were essentially transferring within the same corporate family? That, from a company that would fire 1,200 people barely two weeks after Legere got there. On November 15, 2001, when the firm revealed it would lay off 1,200 people, it also announced a write-off of $3.5 billion and said that it would cut its capital expenditure by over a billion dollars. The house of cards was crumbling. It was only a matter of time before it would crash.

While Winnick was stuffing Legere's pockets, Global Crossing's 401(k) retirement savings plans were going to zero—they were stuffed with Global Crossing stock. (See Table 3.1).

Global Crossing filed for bankruptcy on January 28, 2002—the fourth largest bankruptcy in U.S. history. Investors lost $54 billion, and nearly 10,000 employees lost their jobs. A disgusted W. J. "Billy" Tauzin, a Louisiana Congressman and chairman of the Committee on Energy and Commerce, later said: "By now, this has become a familiar,

Table 3.1 Global Double Crossing
Investors and employees suffered as top dogs at Global
Crossing stuffed their pockets.

Gary Winnick, chairman (Total: $750.8 million)	
Stock sales	$735.0 million
Salary and annual bonuses	$2.8 million
Consulting fees	$7.2 million
Aircraft ownership interest	$2.0 million
Office renovations	$3.8 million
Other directors' stock sales (Total: $582.3 million)	
Abbott Brown, early senior VP	$125.5 million
Joe Clayton, former Frontier CEO	$21.5 million
Dan Cohrs, CFO	$8.7 million
Lodwrick Cook, co-chair	$36.1 million
David Lee, early president and COO	$216.3 million
Barry Porter, early senior VP	$174.2 million
The five CEOs (Total: $104.9 million)	
Combined stock sales	$85.4 million
Salaries and annual bonuses	$19.5 million
Early investors' stock sales (Total: $3.8 billion)	
CIBC World Markets	$1.7 billion
Loews/CNA Financial	$1.6 billion
Ullico	$0.5 billion
Grand Total: $5.2 billion	

Source: Adapted from Julie Creswell with Nomi Prins, "The Emperor of
Greed," *Fortune* magazine, June 24, 2002.

if disturbing story. In the go-go '90s when the irrational exuberance of
the marketplace dictated that stocks only increase in value, meeting
Wall Street's expectations came to be seen as the paramount duty of all
too many corporate executives. But that cannot justify what these firms
seem to have attempted with these swaps, any more than the bizarre
partnerships at Enron or the ginned-up books at WorldCom."[40]

4

BILLIONAIRE VERSUS BILLIONAIRE

Nothing irked Joe Nacchio and Jim Crowe more than being mentioned in the same breath. When *Network World* magazine asked Qwest CEO Nacchio about Crowe's Level 3 Communications, a Qwest rival, Nacchio snorted, "Jim Crowe is a smart guy, but Jim is five years away from having a network. Five years in our world and you might as well be selling pretzels today."[1] Four months later, Crowe hit back with, "Joe is building a car to race in the Indy 500 and we're building a space shuttle to get to the moon."[2]

A former Qwest executive recalls that the rivalry was so intense that at a staff meeting, someone mentioned that Level 3 was becoming a force to reckon with in the wholesale dial-up Internet access business, and within hours it was decided that Qwest would build a million-port dial network to wrest the AOL and Microsoft Network business away from Level 3. This was multimillion-dollar ego-tripping.

In the early days of the telecom gold rush, the two Denver kings, Crowe and Nacchio, presided over some of the biggest and fastest fiber-optic networks in the world. Yet the two men couldn't be more different. They are the Cain and Abel of broadband. While Nacchio is an aggressive street fighter in sharp suits, Crowe is a more understated, though equally competitive, chief executive with warm features and a booming laugh. Crowe, who is credited with turning Level 3 Communications from a late arrival in the bandwidth game into a fearsome bandwidth player, prefers to dress casually in Dockers and Level 3 shirts. He is a voracious reader and is also a do-it-yourself technology nut.

Crowe became a telecom legend long before Nacchio rode into town. In 1996, he sold his first telecommunications start-up, MFS Communications, to Bernie Ebbers' WorldCom for roughly $14 billion. Nacchio, by comparison, rose from the ranks at AT&T and was a relative newcomer to the broadband mafia at that time. In 1996, while Nacchio was busy fighting political battles at AT&T, Crowe had already bought UUNet, one of the first commercial Internet service providers, for $2 billion, assuring his entry into the then-exclusive broadband world.

There's another way in which Crowe and Nacchio are different. After the broadband bubble burst in 2002, Nacchio had to make an ignominious exit from Qwest, while Crowe is still standing—barely, but standing nonetheless.

■ ■ ■

Jim Crowe is the son of Mona and Henry P. Crowe; his father was a decorated World War II hero. He was born in Camp Pendleton, California, on July 12, 1949. He became one of those rare corporate chieftains who can switch from being a chief executive to being a techno-nerd with remarkable ease. He is smooth, well-spoken, and has a wry sense of humor that endears him the most to his financial backers, bankers, and employees. His forthright and direct manner is one of the reasons his company has weathered the telecom storm while others' ships have sunk. True, Crowe got caught up in and perpetuated the broadband hype, but at least he never sugarcoated bad news and was mostly, if not always, upfront about Level 3 with investors.

As a young man, Jim studied mechanical engineering at Rensselaer Polytechnic Institute in Troy, New York. He later attended Pepperdine University in Malibu, California, got his master's degree in business, and ended up working for Morrison Knudsen, a power plant builder from

Boise, Idaho. (The company acquired Raytheon Engineers & Constructors in 2000 and renamed itself Washington Group.) He worked at various locations for the company—from Saratoga, New York, to Washington, D.C., to Boise, Idaho. When denied a promotion in 1986, he left Morrison and teamed up with the rival contract giant Peter Kiewit Sons, based in Omaha, Nebraska.

At Peter Kiewit, Crowe met Walter Scott, the company's chief executive and the man who became largely responsible for Crowe's entry into the broadband world and the attendant wealth and success. With electricity deregulation getting under way, Crowe convinced Scott to invest in some power plants, an investment that proved to be highly profitable. In 1988, with backing from Scott, Crowe started MFS Communications, a competitive local access provider—a precursor to what would be known as competitive local exchange carriers, or CLECs. The business was hardly fashionable at that point in time, but Scott put about $500 million into the company, which went public in 1993.

Through his connection with Scott, Crowe became part of an elite club of high-powered executives who would gather periodically to exchange ideas and corporate war stories. Members of this exclusive yet informal cabal included Warren Buffett and his billionaire friends like Microsoft founder Bill Gates and then Coca-Cola chief executive Don Keough. In 1995, while attending a biannual retreat organized by Buffett in Dublin, Ireland, Scott learned about the Internet from Gates, who was ironically, already late to the game. "Afterwards, I sat down with Bill and talked with him about it. My gut feeling was, if you weren't part of it, you were going to be left behind," Scott said.[3]

By 1996, Crowe believed that the Internet was about to explode and quickly spent a staggering $2 billion to buy UUNet Technologies, then a major Internet service provider. Even before the two operations could be completely merged, a few months later WorldCom came calling, and Crowe sold off MFS and UUNet for $14.3 billion. So what if MFS was still unprofitable! Scott and Crowe walked away with a total of $636 million in WorldCom stock and options.[4] Crowe became chairman of the merged entity, MFS-WorldCom, which would soon gobble up MCI and become the T-rex of the telecom business.

The success of MFS made any executive associated with the company a hot commodity in the post–Telecom Act of 1996 era. Its alumni were everywhere. Among those who jumped ship after MFS was acquired by WorldCom were Royce Holland, who later started Allegiance Telecom; Mike Malaga and Bill Euske, who founded now-defunct NorthPoint Communications; Cindy Schonhaut of ICG Communications; Kirby "Buddy" Pickle of Teligent; and Bob Taylor of Focal Communications. At the height of the broadband bubble, Holland observed: "When I worked at MFS you could fit all the vice presidents in a phone booth, now you need the Yankee stadium," obviously riffing on the high regard with which MFS executives were held during the 1990s boom.[5] Such was the MFS mystique that every CLEC CEO wannabe pretended that he or she was an ex-MFS executive.

But the man who truly became a cult hero for telecom entrepreneurs was Jim Crowe. His reputation skyrocketed after the multibillion-dollar sale of his unprofitable company. The sale of MFS had made him very rich—he sold a small percentage of his WorldCom stock for about $65 million. But at MFS-WorldCom, there was enough room in the spotlight for only one. Ebbers and Crowe both were larger-than-life personalities, and barely a month after the MFS-WorldCom deal closed, Crowe left WorldCom and went back to Kiewit to run its Kiewit Diversified Group, a company that later became Level 3. Along with Crowe went four others who had joined the WorldCom board after the sale.

Thanks to Crowe's newfound stature in the telecom business, Phil Anschutz, the enigmatic billionaire and the real power behind Qwest, soon invited 51-year-old Crowe to join Qwest's board of directors.

At Qwest's board meetings, Crowe observed Qwest chief executive Nacchio's ambition. Nacchio was expanding the company's network and had persuaded companies like GTE, Frontier, and WorldCom to buy half the network's capacity for $3.5 billion, which nearly paid for the cost of building the network. Crowe was impressed and intrigued. He realized the big moneymaking potential of optical networking technology, and it wasn't long before he was conferring with his patron and close friend Walter Scott, who agreed to pony up the cash if Crowe decided to build a brand new company. Scott was betting on a sure thing—after all, MFS had made Peter Kiewit Sons billions. And no less than Bill Gates had assured him that the Internet was a no-lose proposition.

Itching to get back into the game, Crowe quit the Qwest board and started what became Level 3 Communications. The company established a beachhead in Broomfield, Colorado, and 18 former MFS guys high-

tailed it to Broomfield to help Crowe get Level 3 off the ground. Bernie
Ebbers saw this as mutiny, and Nacchio was hopping mad. The tempera-
tures were rising in the Qwest board room as well. "I said that I might
start a company and it might be competitive," Crowe told a *Wall Street
Journal* reporter later.[6] Nacchio's team assumed Crowe was thinking
about building a local phone company, as he had done with MFS.

But nothing had prepared Anschutz and Nacchio for a frontal assault
from Crowe, who started out with $3 billion in the bank, mostly from
Scott, Peter Kiewit Sons, and other private investors. Scott even became
the chairman of the new company. "At the end of the day, I have always
questioned why he would join [Qwest's board]. I'll bet you he learned
something being on our board," Nacchio bristled.[7] He dismissed Level 3
as a copycat. Crowe retorted by claiming that Level 3 would zoom past
Qwest because it had better technology.

Meanwhile, for Omahans it was sweet revenge; many in Omaha
were in a huff over Phil Anschutz's hardball tactics during the Union
Pacific–Southern Pacific Railroad merger several years earlier. Anschutz, a much tougher negotiator than the soft-spoken Omaha folks,
had sold Southern Pacific for a whopping $5.4 billion and still got to
keep the fiber-optic network. With Level 3, Anschutz was finally getting his comeuppance.

Another thing was clear to Omahans: Jim Crowe's unshakeable belief
in the eventual demand for bandwidth. Level 3 executives dismissed
doubters who worried about too much supply and too little demand.
"With all due respect, most of the analysts who say that failed Economics
101," Crowe told *Fortune*.[8] Crowe believed that lower prices would drive
incumbents like MCI, AT&T, and Sprint out of the business.

Level 3 proposed building a 16,000-mile-long fiber network, which
would use the very latest technology to transport Internet traffic between
cities around North America and connect some of the major world capi-
tals. And it would be better than Qwest's network. Crowe was quite vocal
about that, and it was no surprise that he and Nacchio started to take
cheap shots at each other.

In the race to catch up with Qwest, Crowe decided to forgo the traditional initial public offering route, and instead took over a tracking stock
held by Peter Kiewit Sons and listed on the NASDAQ in April 1998.
The reason for this was quite simple. In the halcyon days of the late

1990s, the stock price of a company was seen as the best "calling card" a chief executive could have. At the time, Qwest stock was trading at about $18.60 a share, while WorldCom was trading at $29 a share. The so-called new carriers were the talk of Wall Street. Nacchio, Ebbers, and even lesser mortals like Global Crossing's Gary Winnick were becoming cult heroes, and Crowe and Level 3 simply needed to steal their thunder.

Level 3 started trading under the ticker symbol of LVLT at a split-adjusted $37.12 a share on April Fools' day in 1998. Later that month, Level 3 raised $2 billion in debt in a private placement handled by the investment bank Salomon Smith Barney. Its rising stock price helped Level 3 raise another $834 million in debt in September 1998, again with Salomon as lead investment banker. By the end of 1998, the stock was trading at $43 a share. The stock got another boost in July 1998, when Craig McCaw, the wireless wonder boy and member of the billionaires club, bought $700 million of Level 3's capacity for Nextlink, his newest venture.

Salomon Smith Barney analyst Jack Grubman initiated coverage of Level 3 on January 9, 1999, with an outperform rating. He called Level 3 a great play on bandwidth, which would be scarce. Not surprisingly, he wrote pretty much the same thing about a dozen other companies. In March 1999, Level 3 sold another 25 million shares for $54 a share. But Grubman had already raised his price target on the stock to $70 a share, which was a sign to investors: Buy this stock, and in less than a year make $16 a share profit. Sure enough, that happened—only faster. At the end of March 1999, the stock was trading for $72 a share. By April 1999, Level 3 was being bought and sold at $90 a share, adjusted for splits. Grubman, who had dubbed Level 3 "the Intel inside of telecom," was a star again!

Meanwhile, in July 1998, Level 3 had broken ground on its 16,000-mile U.S. network in Schulenburg, Texas, and would also build another 4,750-mile network in Europe. In order to get on the fast track, Crowe called one of his Omaha buddies, Richard Davidson, head of Union Pacific Railroad, and got permission to build along the railroad's tracks.

Level 3 used about 1,000 people across 20 time zones and completed its network in record time. After all, money was no object. Wall Street, having fallen under the spell woven by Crowe, decided to give him the ultimate platinum card. Over the next four years, Level 3 went out and raised a whopping $13 billion, much of it through junk bonds, which it used to build and expand its network.

"Monopolies offend me. They stifle innovation and lead inevitably to waste. Introducing markets to monopolies is a lot of fun," he would later tell *Forbes*, which quickly dubbed him "Bell Buster" on the cover of the magazine.[9] Bankers loved that kind of stuff. And when Crowe started spouting techno-babble like, "We're watching a change in the whole telecommunications infrastructure that is on the scale of the shift from mainframe computers to the PC,"[10] investors applauded with their dollars. Level 3's founding principles, Crowe would point out time and again, were to increase Internet bandwidth demand by pursuing ever lower prices. "For every 1 percent you drop price, you get a greater than 1 percent increase in demand," he chanted repeatedly. In the early days of the Internet boom, there was hardly anyone who would question Crowe.

The stock continued to rise, and it wasn't long before the growth mutual funds decided to hoard Level 3 shares. By March 2000, Level 3 was trading at $132 a share, giving the company a market capitalization of $46.2 billion, which was more than some old-economy companies like General Motors. Few noticed or cared that the company, in its first two years of existence, had burnt through almost $10.5 billion, which it had raised through debt and stock offerings. While the company had $1.2 billion in sales, it lost $1.45 billion in 2000. Investors dismissed it as the cost of building the network and kept bidding up the stock. Crowe sold about 4 million shares in 2000 and 2001 for about $70 million. He became involved in many community activities and was a local hero.

Level 3 stock made many locals extremely rich. Once Level 3 was carved out of the parent Peter Kiewit Sons, its stock was distributed among employees, who became so rich that Omaha became one of the most expensive places in America. According to the *Omaha World-Herald*, the hometown paper, since the early 1990s homes costing $500,000 and higher had increasingly become part of the Omaha landscape. With Level 3 stock touching $130 a share at one point, there were many millionaires in this city of 390,000.

Awash in Bandwidth

But 2000 was not going to be an easy year for Level 3 or any of its competitors—of whom there were many. One of the more aggressive ones, and a latecomer to the party, was 360networks, a Canadian company that had started life by laying fiber along the tracks of CN Rail in Canada. Originally known as Ledcor Communications, the company was 55 percent owned by Cliff and Dave Lede, brothers who had inherited their stake from their father, Bill, who died in 1980 in a construction accident. Vancouver, British Columbia–based Ledcor morphed into 360networks; the company completed its network in two years and lured Microsoft chief financial officer Greg Maffei to come in as the chief executive. With Maffei's help, 360 went public, raising about $782 million in April 2000. (360networks offered 39-year-old Maffei 62 million shares of the company, at $1.25 a share, and it also lent him $77.5 million to buy those shares. On the day 360networks started trading, Maffei's stake was worth $1 billion.) Meanwhile, the Lede brothers at one point in 2000 were the seventh richest people in Canada, worth a combined $4.6 billion, thanks to their massive holdings in 360networks.

360networks was also using the latest gear available in the market and was willing to play the pricing game, à la Level 3. Most fiber-optic carriers, like Level 3, Qwest, 360networks, and Global Crossing, were so competitive that bandwidth prices started falling drastically. For much of 1998 and 1999, the bandwidth demand had been on an upswing due to several factors:

- Venture capitalists had been putting too much money into dot-coms, which in turn led to a spending orgy.
- Since these dot-coms hosted their web sites at Web-hosting companies such as Exodus Communications, the demand for bandwidth to transport information back and forth from the data centers shot up dramatically.
- At the same time, companies (such as Covad, NorthPoint Communications, and Rhythms NetConnections) that promised to deliver high-speed Internet access to consumers using digital subscriber line technology were beginning to grow quickly.
- But in March 2000, the dot-com bubble popped and venture capitalists stopped funding those companies. That, in turn, led to a slowdown in bandwidth demand from the likes of Exodus. At the same time, DSL companies also started failing because of the level of competition from incumbent players like SBC Communications and Verizon.

- The demand from Fortune 500 companies that many start-ups had counted on didn't emerge, largely because the overall economy began to lose steam.

As a result, bandwidth providers like Level 3 and Qwest saw their sales start to slide. Some, like Global Crossing and Qwest, resorted to "capacity-swap" revenue deals (see Chapter 2), but Level 3 was fighting the price war in the marketplace as best it could. Reality caught up with all of these companies in 2001.

Reality Check

On January 26, 2001, Level 3 announced that its fourth quarter 2000 losses were $552 million, on sales of $433 million. "Demand for our services continues to be strong and we continue to see our revenue grow at a very rapid rate," Crowe told *Bloomberg News*.[11] But apparently that wasn't enough to assuage the fears of investors, who kept pushing the shares down. The whole telecom sector had become suspect, largely because other carriers, such as PSINet and Winstar, said that they were contemplating bankruptcy or cutting jobs to conserve cash.

"Investors are worried, and they're most worried about companies that haven't reached maturity stage. Level 3 falls into this category," Rohit Chopra, an analyst with Deutsche BT Alex Brown, told *Bloomberg News*.[12] Level 3 was trading at around $11 a share in April 2001. It rebounded a little to about $14 a share, but then started to slide again after April 18, 2001, when the company announced its first quarter 2001 results—a loss of $535 million on sales of $449 million. By the summer of 2001, Level 3 decided to lay off 2,000 employees and cut its spending. The bad news would continue through much of 2001.

Fear and Loathing in Omaha

Many who bet on Level 3 have lost big—some their homes and others their livelihoods. *Fortune* magazine estimates that "as much as $20 billion—nearly half of Level 3's stock market value at its peak—was lost, on paper at any rate, by Omaha shareholders."[13]

"The whole of Omaha was pretty heavy into Level 3. There was just a tremendous amount of confidence in Walter Scott, Jim Crowe (Level 3's chief executive) and the Level 3 concept," Steve Shanahan, the owner of the Shadow Ridge Country Club in Omaha, told the *Rocky Mountain News*.[14] "I know of a couple of stories where guys sold businesses or completely eliminated their net worth to get into Level 3 in a big way," said Ron Carson, owner of Carson Feltz Retirement Planning Inc., in an interview with the *Omaha World-Herald*.[15] As for Crowe, he did sell about $60 million worth of stock in 2000. Ironically, even now, no one is publicly griping about their losses. Crowe still owns about 10 million shares of Level 3.

Despite the downturn, locals still have a high regard for Crowe and his track record. In September 2002, when the clouds of despair were gathering over Level 3, Crowe's backers, Peter Kiewit Sons and two of its directors, Scott and William Grewcock (both directors at Level 3 as well), bought warrants worth $32 million. They can turn these warrants into common stock of Level 3 shares at $8 a share, anytime before June 30, 2009. Investors viewed this as a vote of confidence.

This investment came a couple of months after Warren Buffett, the sage of Omaha and the second richest man on the planet, decided to invest $100 million in the network carrier company. His blessings led to the formation of a consortium that invested about $500 million in Level 3. The investment came after a conversation between Crowe, Scott, and his longtime friend Buffett. Not wanting to be chastised in the media for his close relationship with Scott, Buffett suggested that O. Mason Hawkins, the chief executive and chairman of Southeastern Asset Management of Memphis, Tennessee, conduct negotiations with Crowe. Longleaf, Legg Mason of Baltimore, and Buffett's Berkshire Hathaway were members of the consortium. The investment boosted the sagging fortunes of Level 3. "Liquid resources and strong financial backing are scarce and valuable assets in today's telecommunications world," said Buffett in a prepared statement at the time of the investment. "Level 3 has both. Coupled with the management of Walter Scott and Jim Crowe, in whom I have great confidence, Level 3 is well equipped to seize important opportunities that are likely to develop in the communications industry."

It was the best PR no money can buy. Level 3 wasted no time and put the news in bold type on its web site. But the media outlets questioned the investment—after all, Buffett had trumpeted his refusal to invest in technology stocks for years, and many wondered why Buffett invested in Level 3 bonds and not in Level 3 stock, which had been hovering at all-time lows. "Investors should think twice before following Buffett's lead. So far, the scenario he envisions for the fiber-optics business doesn't seem to be playing out on a large scale," noted BusinessWeek Online in its "Street Wise" column.[16]

Things got tougher for Level 3 after the Buffett investment. The credit rating agency Standard & Poor's downgraded Level 3's debt in early August 2002, and then a few days later, the Securities and Exchange Commission announced that it was looking into Level 3's financial statements. The SEC was contemplating a move that could clean out $232 million in Level 3's noncash profits from the past four quarters. Apparently, the company was swapping its stock for outstanding debt. It issued new stock for about $364 million in debt, and then recorded $232 million in noncash gains on the deals.

Such news has diminished the aura around the company since Buffett's investment. Crowe's credibility also took a hit when, in August 2002, the *Wall Street Journal* reported that Crowe and Scott had received Qwest shares at the time of Qwest's initial public offering. The news came a month after Crowe had denied similar charges made by a former Salomon broker. In an earnings call with investors, Crowe said: "I have never purchased, owned, or sold any shares of any of the companies that have been identified in press reports in connection with the alleged IPO purchases, which I understand to be Rhythms NetConnections, Alamosa Holdings, Radware, interWAVE Communications, Focal Communications, US LEC, Global Crossing, Imstat Fiber Networks, and KPNQwest. I have never sought to purchase shares in hot IPOs in return for giving business to any investment bank." Chief executives of rivals like Qwest and WorldCom are now under fire—and in some cases, in court—for precisely those accusations.[17]

Crowe may not have indulged in the wrongdoings of his rivals, but he no doubt hyped the bandwidth bubble. On April 11, 2002, Crowe was quoted in the *Wall Street Journal* as saying, "In six months the [fiber-optic capacity] market will change from surplus to shortage." Nine months previous to that, he had made a similar claim in the *Rocky Mountain News*: "We have some excess capacity today but someplace over the next three to nine, six to nine months, we're going to run out of actual capacity again." The connection, made by *Rocky Mountain News* business editor Rob Reuteman in his April 13, 2002 column, is yet another example of a CEO making wild predictions that stretch reality.[18]

Despite all its broadband pretensions, Level 3 is now swallowing up software companies and buying sales, just to stay out of bankruptcy. With its CorpSoft and Software Spectrum acquisitions, Level 3 is now as much a software reseller as a telecom company. Even today, despite its pretensions of being a broadband leader, a large chunk of its revenue comes from its dial-up operations. In December 2002, the company paid $250 million and bought most of the assets of failed network provider Genuity. It is now a powerhouse in the dial-up business. The company, which had about $7.5 billion in debt at the end of 2001, has lost around $5.5 billion in its brief history. In comparison, Qwest has a net debt of $20 billion and Global Crossing's total borrowing at the time of its bankruptcy stood at $7.6 billion.[19]

David Gross, an analyst with Communications Industry Researchers, believes the company is surviving the telecom recession by not being a telecom provider. It has about $1.5 billion in the bank, enough to keep it going until the industry turns around. Given that over 90 percent of Level 3's debt matures, or comes due, in 2007 or later, Crowe might finally get to beat longtime rival Joe Nacchio to become the king of the (telecom) hill.

5

THE ATTACK OF THE CLONES

Jimmy Luu could not contain himself—for $44,000, he had won the bid to own the crooked "E" logo at an open auction in late September 2002.[1] His boss, an owner of three-store computer chain MicroCache, would be pleased. The auction of Enron's remnants had attracted thousands to a Houston hotel. Many showed up at the Radisson Astrodome around 5 A.M., four hours before the auction; about 1,000 got in, and the rest stood in lines that stretched outside the hotel. Another 12,000 signed up for the auction over the Internet. On sale were stress balls, mugs, and an air hockey table, and plasma televisions that went for $8,000 a pop.[2] Even ex-employees showed up. Brian Cruver, a former Enron employee and author of Anatomy of Greed: The Unshredded Truth from an Enron Insider, *showed up, hoping to buy back his old chair. Apparently, despite a glut of Enron books on the market, he had made enough to bid for his throne. "I'm here to buy my old chair. It's the most comfortable chair I've ever sat in and I want it back," he told the Associated Press.[3] It was a macabre fascination that brought them to the sale with open checkbooks. Enron, which wanted to make a market in bandwidth using an electronic version of auctions, was getting its due in a familiar fashion.*

The Great Bandwidth Bazaar

In early 2000, Jeffrey Skilling and Ken Lay decided they needed to put on a new show. The CEO and chairman of Enron, a Houston-based energy trader, had grown tired of doing the oil and gas routine. Hoping to cash in on the great broadband boom, they cast themselves as the new messiahs of broadband, the pashas of a vast new bandwidth bazaar that would make every Enron investor wildly rich.

Enron was one of the many outsiders who jumped into the broadband business. Its foray and attempts to muscle in on the broadband business were purely opportunistic, for personal gain and as a means to sprinkle some broadband pixie dust on their old-economy stocks. In the process, Enron ruined a once thriving market by starting a price war, which has brought the telecom industry to its knees.

Enron's entry into the broadband business was pure greed and a series of coincidences. In 1997, Enron bought Portland General Electric, a small Oregon utility, for about $3.1 billion in stock and assumed debt. PGE, as it was then known, had a division called First Point Communications, which was building a fiber-optic network. First Point was under the stewardship of Joe Hirko, who previously served as chief financial officer of PGE. After the acquisition, First Point's name was changed to Enron Communications. (It would eventually be reinvented as Enron Broadband Services.)

The company was making money from its network, and Hirko felt there was big potential if the network could go national. A few weeks after Enron bought PGE, Hirko was hopeful that he would get a positive response from the senior management. Hirko and David Harrison, one of the architects of First Point, flew to Houston and made a pitch for taking their network national. Skilling reportedly couldn't stifle the yawns as the two talked about optics, networks, and routers.

Hirko proposed that the company get deeper into the broadband business. They wanted to expand their little network to span 1,500 miles for about $50 million. Hirko explained that the company could trade bandwidth like any other commodity.

The buying and selling of bandwidth at the time was hobbled by 20-year contracts, and the pricing terms were very rigid. At the same time, more capacity was coming online, and it meant that prices were going to fall. To Hirko it made sense that, like Enron had done before in the oil, gas, and electricity markets, it should become a disrupter in the bandwidth markets as well. Now *that* perked up Skilling's ears—he loved to trade and wanted Enron to be a big player in all sorts of markets, even bandwidth. "Jeff just wasn't interested. Jeff saw it as an investment opportunity, not a core business. He made it clear we couldn't be a drain on Enron—and we weren't," said Hirko later.[4] Eventually Hirko got his wish.

But Enron's bandwidth trading idea was not original. Band-X, a London-based telecom broker, had been in business for almost two years before Skilling and the rest of his acolytes got onto the bandwidth trading bandwagon. In the summer of 1996, Richard Elliott, then an analyst with Kleinwort Benson, an investment bank, and Marcus de Ferranti, a former Harrier pilot working for the British government on guided missile systems, came up with the idea of starting a neutral marketplace to facilitate the selling of voice minutes, related infrastructure such as fiber, and bandwidth. Their initial effort was amateurish: Their web site was a small bulletin board with two big oval buttons—"Buyers Enter Here" and "Sellers Enter Here." Nevertheless, the firm eventually became an active player in the buying and selling of anything telecom, and has since expanded to three continents. In late 1998, Elliott got a visit from some folks from Enron. "They were interested in the market; we were clear leaders in it. We trusted their U.K. management—in retrospect, somewhat naively," said Elliott.

Elliott recalls chuckling at the grandiose statements Enron was making, claiming that it was the first company to do this or do that. "My attitude was one of bemusement, really. Some of the things they were saying they could do we'd already had a go at or examined in detail (and passed on that)," Elliott said. "They certainly played a part in encouraging carriers to think again about the way they held and accounted for inventory."

Elliott, an analyst in a former life, had deduced that there was an oversupply in the market, largely because the number of facility-based providers had ballooned to around 500 worldwide, with about 300 in the United States alone. This included long-haul network operators such as Level 3, Qwest, IXC, Williams, and KPNQwest. The total fiber miles had shot up drastically, and newcomers like 360networks and Aerie Networks were looking to muscle in on the business. What Elliott was not prepared for was the voracity with which Enron would compete in the broadband market. Enron spent around $2 billion on the broadband project, but that was a small price to pay, as the company's stock zoomed. Wall Street bought Enron's grand vision of $700 billion in bandwidth trading volumes that would run through Enron's grand bandwidth bazaar.

Around that time, the broadband stocks, including those of carriers such as Global Crossing, Qwest Communications, and Level 3, were going through the roof. Jim Crowe and Gary Winnick were gracing the covers of magazines such as *Forbes*. They were the *IT* companies of the late 1990s, even bigger than the dot-coms. In late 1999, with fiber-optic stocks going to the moon, Lay and disgraced former chief executive Jeffrey Skilling suddenly figured the potential upside of broadband offered for Enron stock.

In December 1999, Enron announced its first bandwidth trade, and on January 20, 2000, in a stock analysts meeting, Lay and Skilling waxed eloquent about broadband. Lay predicted that broadband trading would dwarf Enron's traditional gas and power trading business. A few minutes later, Scott McNealy, the toothy billionaire chief executive of Sun Microsystems, came on stage and pumped up the hype a wee bit more. What did he have to lose? After all, Enron had promised to buy $350 million worth of new Sun servers.

Enron stock jumped from the low-$40s to $70 a share, adding a cool $21 billion to its market valuation. "Enron came in with a disruptive price, undercutting the current prices by almost half," recalled Brent Wilkins, managing director of Cantor Fitzgerald Telecom Services, a new moniker of Chapel Hill Broadband, a telecom brokerage where he was the president. Enron's plan was to make the wholesale bandwidth market more liquid. Being traders, Enron folks knew how to get the market started but had no clue about the subtle nuances of the telecom industry.

While it may be fashionable to blame Enron for everything these days, it did cause a serious problem in the telecom and bandwidth business. "It was like you mortgage your house, max out your credit cards, and then lose your job at the same time," said Wilkins, who remembers not being able to figure out how Enron was doing business or was making money. "I couldn't see the volume of sales and I couldn't understand how they would deliver on what they promised."

For instance, Enron could sell the bandwidth contracts, but could never actually deliver the service it promised because it had no access to the local networks, or customer premises. All the price cuts were arbitrary. Not satisfied with the price undercutting, Enron decided to make the actual pricing data and indices available. This resulted in corporations (the large portion of end-users) demanding price cuts from their carriers. Revenues and any notion of profits nose-dived. Enron and other energy companies did have one positive impact—they ignited the demand for equipment, which led to another bubble, this time in equipment stock—but that would be later. In the summer of 2000, emboldened by the stock market's enthusiastic reception, Skilling, Lay, and others got even more ambitious.

To realize their plans, they had to get rid of the old First Point people and replace them with the Houston trader jocks. By June 2000, Hirko was eased out and so were others. For getting Enron into the bandwidth game, Hirko was paid handsomely. Between January 2000 and May 2000, he sold $35 million worth of stock, enough for him to enjoy watching Oprah for the rest of his life.

Enter Ken Rice, the new chief executive of Enron Broadband Services. Author and journalist Robert Bryce describes him as "flashy, charming, gregarious, and a 6-year-old in a 40-year-old's body" in his best-selling book, *Pipe Dreams: Greed, Ego, and the Death of Enron.*

A former gas pipeline executive, Rice started his career with Inter-North (a company that was one-half of Enron) in 1980. An electrical engineer from the University of Nebraska, Rice got his MBA from Creighton University. When InterNorth merged with Houston Natural Gas, the combined company was named Enron, and Rice became a salesman for the natural gas giant, one of the very best. He was the elephant hunter for the company, and that was what endeared him to Jeff Skilling. Every big kill meant more money in Rice's pocket.

He made so much money that he did not have to show up at work. He could spend hours with his high performance cars, Ferraris—two $160,000 360s and one $200,000 550 Maranello—or his BMW motorcycles. And then he was named the head of Enron Broadband Services. Not that it really affected his lifestyle much. Bryce, in his book, points out that Rice would show up maybe three or four times a week to work, often in cowboy boots and jeans, and perhaps watch cartoons during meetings.[5] At the time, Rice was more known inside Enron for his affair with one Amanda Martin, whom he would later marry. The couple, though married at the time to other people, were quite blasé about their affair and would shock their co-workers. It wasn't unusual to find them in, to say it politely, compromising positions.[6]

Still, being Jeff Skilling's buddy meant that Rice had the proverbial pass to do just about anything. One of his first acts as the head of EBS was to buy two motorcycles from ultra-exclusive bike manufacturer Confederate Motorcycles.[7] The man was clueless about broadband business, and so was his number two in command, Kevin Hannon. Insiders described the president of Enron Broadband Services as arrogant, remote, and Mr. Know-it-all. Rice and Hannon were both Skilling cronies, and that alone was enough to ensure that Enron Broadband was going to be a colossal mess.

The grand plan was to build two dozen pooling points across the globe. These pooling points, which would use sophisticated equipment from companies such as Sycamore Networks, would act as hubs. Enron would buy and sell capacity on the Internet so that corporations and other carriers could get reliable network connections. In theory it was a sound plan, but in reality it was far from real. In order to achieve something of this magnitude, what Enron needed was the ability to touch customer premises or other carriers' networks. And in order to do that, Enron had to act humble and friendly—which it didn't. It came into the business saying, "You, UUNet, better get ready, because we are going to eat you for lunch and Qwest for breakfast."

And there was another problem—the technology. At the time, and even now, no one had built equipment that could help provide instant bandwidth on demand. Not even Sycamore, Enron's vendor of choice at the time. "We don't think it works. We've heard the pitch. We asked questions. We didn't get any answers," said Stan Woodward, an executive with Yahoo! Broadcast.[8] Yahoo! would have been a perfect customer for Enron.

Enron's arrogance stemmed from the fact that, about a decade earlier, Enron had successfully launched and dominated the natural gas and then the energy trading businesses. The concept was that a commodity is a commodity, be it electricity or bandwidth. One small thing they overlooked—unlike natural gas or electricity, bandwidth was not a necessity, and thus not a perishable commodity. Rarely do thousands of companies need gigabit connections at the same time, unlike electricity and natural gas, which are high-volume, low-margin businesses.

Despite its insistence, Enron was having difficulty finding either customers or partners. In its arrogance, the company announced that it would publish the bandwidth prices on the Web for all to see, which created a problem for other broadband companies that were trying to stay solvent. Enron's decision to publish discounted prices only accelerated the price wars that had broken out among hundreds of carriers. It alienated the company further from potential partners.

While Enron Broadband was spending hundreds of millions of dollars, it was finding it hard to attract customers. In addition, EBS was made to hire hundreds of employees from divisions that had been eliminated or people who were being downsized. Still, there was the appearance of success—trades were happening, and revenues, however fake, were being generated. One trick to bring revenues up was to sell circuits to another company, say Company Z, which would in turn trade the same circuit over to another company, which would turn around and sell it back to Enron. So all three could book revenues, after all they had sold something—so what if no cash had traded hands. Since these were days of broadband madness, no analyst bothered to ask the right questions. Like Global Crossing or 360networks, no one expected EBS to start making money that quickly. And this gave Rice and his sidekick, Hannon, space to goof around.

After Rice took over as the chief executive, in July 2000, Enron Broadband announced with huge fanfare a sweeping agreement with Blockbuster, the video rental people. The two companies proposed to offer video-on-demand—not entirely a new concept, at least for the subscribers of cable television services. It was simply a ploy to make Enron seem cool and drive up the stock price—and it did. Stock was in the $80s at the time of the announcement.

Those who forget history are likely to pay the price, and most forgot that this whole nirvana of video-on-demand had cost Time Warner nearly $5 billion earlier in the decade as it tried to develop an intelligent cable network. Investors, illogical as ever, kept driving the Enron stock higher, which hit $90 a share in August 2000. A month later, in September 2000, Enron announced that it would spend about $2 billion in the broadband operation. Enron said it was taking over the bandwidth market, and everyone believed them.

Kevin Hannon, president and chief operating officer of Enron Broadband Services, in an interview[9] claimed that Enron had completed at least 300 trades—all imaginary, as we would later find out—since it had started trading bandwidth in the fall of 1999. Kevin always had a great imagination. A month earlier he had told me how Enron would become a major player because of its experience building similar trading platforms for the natural gas and electricity markets. While some of us did not believe this one-time trader, apparently many in Houston did!

In Houston, for the longest time it was said that as goes Enron, so goes Houston. Enron's bandwidth trading mantra was not lost on its counterparts in the oil, gas, and energy business, who all decided if it was good enough for Enron, then it was good enough for the rest of them. Enron, the most powerful company in the energy belt, was being rewarded by investors for its move into broadband and bandwidth trading. Hoping to share the wealth, Williams, El Paso Energy, and several others announced their trading ambitions. What they did not know was that Enron was a house of cards, and that as it fell, it would take many of them down as well.

To enter the business, all these staid old boring companies had to do was run fiber-optic cables through their pipelines and get into the bandwidth business, and their stock would go shooting to the stars. In the summer of 2000, Dynegy stock was trading in the mid-$30s; by the end of the year the stock had nearly doubled. Why? Because the company spent $151 million in cash and stock to buy Extant, a data communications company, and establish Dynegy Global Communications. DCG was going to build a nationwide, optically switched network that would consist of approximately 16,000 route miles and more than 40 points of presence, the company said.

It's easy to see why these companies wanted a new market to pursue. The Yankee Group, a research group, estimated that the energy market grew less than 2 percent annually, while the telecom sector was expanding at roughly 8 to 10 percent annually. Aside from the better forecast, telecom also had better margins than the pure energy business. "With the energy business, the regulated rates of return are in the 12 to 14 percent range. In telecom there are no regulations on the transport side, so our returns are more in the 20 to 30 percent range," said Frank Semple, president of Williams Network, a division of the Williams Communications Group, justifying his company's entry into the communications business.[10]

By January 2001, there were at least 50 energy companies which had some sort of telecom divisions, according to Edison Electric Institute, an energy industry–funded trade group. Europeans were equally bullish, and obviously win the coolest-names contest for coming up with monikers such as 186K (after the speed of light), 51 degrees and UrBand, a joint venture between 186K and Thames Water of London.

The entry of companies like Enron and other utility companies into the broadband business was like pouring fuel on the fire. The demand for basic equipment needed, such as optical switches, transport systems, and fiber, went skyrocketing to about $105 billion in North America in 2001. Typically, a single energy company would spend about $1 to $1.5 billion dollars on their networks, and about two-thirds of that would go to companies such as Sycamore, Lucent, Nortel, Corning, Ciena, Cisco, and Juniper, all makers of a variety of telecom and broadband gear. More networks meant more orders (and revenues) for these companies, which resulted in these stocks going ever higher. In short, this ill-planned move added more air to the equipment bubble, and also helped accelerate the demise of pure-play telecoms such as Global Crossing, which was selling more for less, thanks to the competition from energy companies.

But back to the clueless of Houston. On January 25, 2001, the slippery tongue of Skilling painted yet another rosy picture for the analysts. He claimed that Enron Broadband was going to be a cash cow. While he was telling these half-truths, EBS's 2,000 employees had trouble finding customers or revenues. Or Ken Rice! Skilling was going to make broadband services the focus of his presentation to the analysts, and he was going to forecast that Enron's share price would go to $126 a share (from $82 a share). It would have been nice if the chief of Enron Broadband was around to back him up. They tried for two days to find Rice, without much luck. He was out racing cars. Somehow they tracked him down and prepped him enough to get him through a speech and a 64-slide PowerPoint presentation.[11]

He waxed eloquent about data storage market, bandwidth trading, and a wonderful future that should be worth about $40 billion in market capitalization. Okay, he didn't say that, but you get the picture, even though, in reality, the firm was trying to sell its single hard asset: an 18,000-mile fiber-optic data network. "There wasn't a whole lot of connection between what the management said and what we knew," said former EBS trader Dixie Yeck, in an interview with the *Houston Chronicle*.[12] But Rice had nothing to worry about, for, about a month prior to the meeting with the analysts, Rice had sold 45,000 shares of Enron for a profit of $2.6 million. On the day of the meeting, he sold another 3,000 shares for a total profit of $169,445.

But since the guys at El Paso Energy did not get the memo, in what was "catching up with the Lays," they announced that El Paso was going to get a complete fiber-optic makeover. The Houston-based El Paso Energy had been in business since 1928 when a Houston attorney, Paul Kayser, formed a small company, El Paso Natural Gas, to supply fuel to the city of El Paso, Texas.[13] The company steadily grew to see $57.5 billion in annual sales in 2001, and it also dipped its toes in the broadband business. But for some odd reason, it decided to throw caution to the winds and get aggressive with its broadband strategy.

In February 2001, the company announced that its subsidiary, El Paso Global Networks, would spend $2 billion on a 34-city optical fiber network and would spend about $1 billion on Cisco Systems' networking gear. El Paso was clearly late to the game. There were at least two dozen national networks already under development at the time. Still, the company defended its late arrival: "This business has more potential than any other we've been involved with," Greg Jenkins, the El Paso executive in charge of the fiber-optic business, told *Petroleum Finance Week*.[14] The announcement was a big, fancy event, attended by most Wall Street analysts as well as Mike Volpi, Cisco Systems' rising star and chief strategy officer. Cisco CEO John Chambers touted the new venture in a video presentation.

At the time, according to analyst reports, the company projected it would have 15 percent of the bandwidth trading market and bring in $4 billion, while a 2 percent share of the long-haul transport business would add another billion or so to its revenue. Another couple of billion dollars would come from transmission of data on the intracity networks. "The opportunity to extend our merchant platform to the telecommunications industry offers significant value creation potential for our shareholders," boasted El Paso Chairman William Wise.[15] "We expect this business to generate positive operating cash flow by the end of 2003, but more importantly, we firmly believe that El Paso Global Networks has created $7 billion to $10 billion of current value based upon opportunities that have already been identified."

Funny how Wise was speaking like Skilling, and Wall Street was still buying it. My theory is that the power and energy analysts on Wall Street were so starved for business (well, compared to Jack Grubman) that they believed everything they were told. Wall Street chimed in with its hyperbole; they had bought the hype. One analyst from UBS Warburg, a New York investment bank, wrote in a report following the launch, "We are now willing to slap on some value to El Paso's Global Networks division. We have no qualm about EPG's analysis, but would simply like to remain highly conservative at this juncture." The bank put the value of the global transmission at about $8 a share and raised its target price to $92 a share. At the time, the stock was trading at about $60 a share.

What's more incredible is that as El Paso touted its new plan and as Wall Street sang its praises, the dark clouds were gathering over the telecom valley, and at least four companies, including NorthPoint Communications, had filed for bankruptcy. The carriers' business model was losing traction with even the most ardent believers on Wall Street. El Paso stock peaked on February 21, 2001, hitting $74.50 a share. It never got to that $92 a share UBS Warburg was talking about.

Well, blame it on Enron—or Enron Broadband, to be more accurate. In March 2001, the bottom fell out. Despite all the positive spin Skilling, Rice, and Hannon tried to put on it, in reality, Enron Broadband was not working, and whatever revenue growth was being reported was actually a figment of the imagination of Andrew Fastow, Enron's chief financial officer.

According to SEC filings,[16] in June 2000 Enron Broadband sold dark fiber, or an unused portion of its network, for about $100 million to LJM2 Co-Investment, one of several partnerships set up by Fastow that would eventually land him in handcuffs. LJM2 paid $30 million up front in cash, and $70 million in an interest-bearing note. This allowed Enron to record about $67 million in pre-tax revenue. Six months later, LJM2 sold that fiber to others for $40 million. However, since Enron had helped find those buyers, it got a finder's fee of $20.3 million. The very same month, the rest of the fiber was sold for $113 million to another partnership created by Fastow. The details of these deals would not come out until later, when it would be too late. In March 2001, the company announced that Enron Broadband had losses of $64 million in 2000 on sales of $408 million. Trading operations never took off, and if that was not enough, outsiders had openly started talking about this scam.

Then, on March 21, 2001, Blockbuster announced that it was calling it quits on its much vaunted video-on-demand venture. The minute the Blockbuster news hit the wires, the bottom fell out under EBS. Two hundred and fifty employees were fired. Enron stock plummeted to below $60 and kept falling. It was a far cry from the $126 a share predicted by Skilling a few months earlier. By April 2001, the company had secretly started selling millions of dollars in computer equipment and telecom equipment. Its attempts to sell its 18,000-mile network were not panning out as well.

Blockbuster and Enron's video-on-demand project was nothing but a sham orchestrated by Enron. "We're still a big believer in digital content delivery. But a couple of things have obviously changed [since last summer]. One is that a PC-based model and standard content delivery—there's just really no revenues in that. You have to focus on recognizable, high-quality content. That's what's going to stimulate connections," Rice told *Red Herring*.[17] Despite all Rice's spin, no one bought the crap EBS was feeding the press.

Eight months earlier, in July 2000, Enron Chairman Kenneth Lay had described the new venture as the "killer app for the entertainment industry."[18] Blockbuster Chairman John Antioco chimed in and described the venture as the "ultimate bricks-clicks-and-flicks strategy."[19] Still, despite these outward displays of confidence, the company had no idea if it would make any money on this project.

First it would have to spend billions developing the infrastructure that would allow Enron to deliver the movies to end-consumers. It would have to buy tons of new gear like storage systems, servers, and routers. It would need DSL lines at the very least, and of course it would need customers. Published accounts show that head honchos at Enron Broadband, including Kevin Hannon,[20] knew that the company would lose about $347 million on sales of $52 million if there were 2 million paying subscribers. Losses would increase to $3.4 billion if Enron got 50 million subscribers.[21]

Despite public posturing, Enron barely established some pilot projects in Portland, Seattle, and Salt Lake City, streaming movies to a few dozen apartments from servers set up in the basements of these building, according to Keith Cooley. "In a previous life I founded a [telecom] company called OnFiber/homeFiber, providing fiber to the homes in Palo Alto and elsewhere, and we were negotiating with Enron about being part of their streaming video initiative with Blockbuster," said Cooley, and added: "We were not able to close the deal and I was very suspicious of its extravagant claims." When Enron failed to deliver, Cooley managed to get the whole picture from some friends about the Enron program.

The real purpose of the Blockbuster announcement apparently was not to deliver the service, but to help Enron grow its revenues. During the fourth quarter of 2000 and the first quarter of 2001, Enron claimed $110.9 million from the Blockbuster deal. Here is how they did it.[22]

- First, Andy Fastow set up a partnership called Braveheart, in an obvious homage to the Mel Gibson movie.
- Enron snookered CIBC World Markets, the investment banking arm of Canadian Imperial Bank of Commerce in Toronto, into putting $115.2 million into Braveheart.
- In exchange, CIBC got to keep all earnings from Enron's share of the Blockbuster venture for the first 10 years.
- Enron got to book revenue and tell the world its broadband business was off to a rocking start.

By the second quarter of 2001, things had gone from bad to worse. The company lost $109 million in the quarter. "It's like someone turned off the light switch. Revenue opportunities have just dried up," Skilling told investors in a conference call in July 2001.[23] The stock had sunk to $49.50 a share, which prompted Skilling to lament, "Everything has been taken out of our stock for the bandwidth business. We are probably getting a negative impact on the stock, and I don't think that's right."[24]

It is clear that Enron executives and directors, in their greed, were trying to drive up the stock price and, in the process, destroying the broadband business. Though they were not alone, they certainly were the most clueless, and greedier than perhaps even Winnick. All together, Enron insiders sold about $924 million of company stock in 2000 and 2001. Rice, who had taken control of Enron Broadband Services once Hirko quit, sold more than a million Enron shares for a total of $70 million in 2000 and the first half of 2001, according to regulatory and court filings. He quit Enron in August 2001.

In the halcyon days of the bubble, Rice predicted that one day bandwidth trading and natural gas trading would be equal. That day came sooner than he expected when Enron went belly-up. In the end, the broadband business became the noose around Enron's neck. Rice, however, has reemerged. He is using his Asian connections and has started International SynerG Communications Holdings Ltd.—he is back in the broadband business.

Things have not gone too well for El Paso Energy—the company is now trading for $8 a share, less than a tenth of what it was supposed to be worth. El Paso's current market capitalization, $4 billion, is less than a tenth of its peak market cap of around $44 billion. El Paso is like many others that got into the business, spent close to billions on new broadband ventures, and then retreated to lick their wounds. Like many others, they should have stuck to their old-economy knitting.

PART II

THE MILE
NIGH CLUB

6

FRESH PRINCE OF HOT AIR

In August 1996, Alex Mandl, on the cusp of heading AT&T, gave up his private jet and multimillion-dollar salary to join a small, unknown start-up company called Associated Communications. It could be said now that Mandl's decision initiated the big bubble in the fixed-wireless sector of the telecommunications industry that cost investors billions of dollars by the time the bubble burst.

Associated Communications, which would later rename itself Teligent, owned licenses for obscure radio frequencies over which it promised delivery of broadband access, video, and voice services, which until then had been transmitted through optical or copper wires. The technology was called fixed wireless, and no one had been able to figure out how to make real money off of it.

Mandl bet his career that Teligent would emerge a leader in the race to offer voice and high-speed Internet access to businesses. Then 52, it seemed he was just a few years away from succeeding Robert Allen, then CEO of AT&T, at the time the biggest and most powerful telecom company in America. The dough-faced and sometimes gruff but well-liked executive had orchestrated a merger with McCaw Cellular that created AT&T Wireless and made Ma Bell a powerhouse in the cellular business.

Like many senior executives at large companies, Mandl wanted his day in the sun—he wanted to be known as a company builder, not just a worker bee all his life. But what made Mandl quit was the

Telecommunications Act of 1996. The first major overhaul of the telecom industry since the breakup of AT&T in 1984, the Act was expected to open telecommunications markets to competition and
provide major changes in laws affecting cable TV, telecommunications, and the Internet. "At AT&T, Alex was quite aware of the Telecom Act and had a calendar in his office that counted down the days
to the Telecom Act—he saw it as an opportunity for AT&T," said Tom
Evslin, chief executive officer and founder of ITXC Corporation and a
former colleague of Mandl.

Mandl rightly believed the Act opened up opportunities to build another AT&T and perhaps get richer in the process. So, when Teligent's
mysterious backers, the Berkman family—billionaires from Pennsylvania—and Raj and Neera Singh, a telecom power couple from Virginia,
approached him to head up operations, he quit AT&T for the promise of
something brighter. "I had a pretty good job. The odds were, I would
succeed [AT&T chairman and CEO] Bob Allen. But I basically woke up
one morning and said, This opportunity is so attractive that if I don't
take it, I probably will never do this kind of thing," said Mandl.[1]

But little did he know that he had bought into a dud. In the coming
years, the company's networks would eventually be built on equipment
that could at best only send voice traffic. There would be no broadband.
A former engineer said that for the longest time even Teligent's own offices got Internet access using the Baby Bells' lines rather than its own
much ballyhooed wireless technology. But the drama would unfold
slowly in the future. The successful and worldly Mandl had just become
a Teligent pawn and, in the end, one of the most tragic figures of the
broadband bubble—one of many—who would lose not just his money
but possibly his reputation.

Austrian-born Mandl moved to the United States in the early 1960s to
attend college. He earned a B.A. degree in economics from Willamette
University in Oregon and an M.B.A. from the University of California at
Berkeley. He began his business career in 1969 at Boise Cascade Corporation as a merger and acquisitions analyst. For the next 11 years he held
various financial positions, including a stint as the company's director of
international finance and treasury functions. In 1980, Mandl joined
Seaboard Coast Line Industries, a $4 billion transportation company, as
senior vice president of finance and corporate planning. After Seaboard

merged with Chessie Systems to form CSX Corporation, he moved to the new parent company as senior vice president in charge of corporate development, human resources, and chief information officer. In addition, he had operating responsibility for three transportation and information technology subsidiaries. In 1988, he was appointed chairman and CEO of Sea-Land Service.

Mandl joined AT&T as its chief financial officer in 1991. It is hard to guess why he moved to AT&T, apart from the most obvious reason: AT&T was a $63 billion (sales) operation, thanks to its near monopoly in the long distance voice business. It was not a company; it was more like a small European monarchy. Mandl's pedigree and Euro-sophistication took him far. Many describe him as extremely polished, chivalrous, and courteous. "He is one of those guys for whom you would walk on hot coals," said Keith Kaczmarek, former Teligent vice president. At AT&T, Mandl oversaw the company's long distance services business, as well as the company's wireless services, online services, multimedia services, and credit cards operations, before being named the president and chief operating officer of the company in 1995. He soon became a board member of the Warner-Lambert Company, Carnegie Hall, and AT&T Universal Card Services. In January 1994, Vice President Al Gore appointed Mandl to the Advisory Council on the National Information Infrastructure, a body that was created during the Clinton administration to help out on all things broadband.

Despite being one of the top guys at AT&T, there was an outside chance that Mandl would be passed over for the top job, and those close to him felt this was something that nagged at him. "I think it became quite clear to Alex that he would not succeed to the top spot because he was not an AT&T lifer. He was very reserved and a very private person, not someone who would share his feelings over a glass of beer. Perhaps that is why, despite being an insider, he looked like an outsider," said Evslin. But if Ma Bell didn't shower him with love, others were willing to show their appreciation for his talent with millions of dollars.

The Shadowy Backers

The Berkmans owned the Bala Cynwyd, Pennsylvania–based Associated Group. The powerhouse behind the company was the late Jack Berkman, a Harvard-educated attorney turned entrepreneur, whose late wife Lillian was also an active participant in the company before she passed away.[2] For as many as 60 years, the Berkmans had smartly bet on new technologies like television, cable television, and cellular phones. Their strategy was to invest in licenses for new technologies in small markets, bundle them together, and flip them to a buyer. In the early 1970s, the Berkmans had acquired several small cable companies, rolled them up, and sold them to John Malone's Tele-Communications, Inc., which was later sold to AT&T for billions of dollars.

In the early 1990s, the Berkmans' son, Myles, along with his two sons, Billy and David, took control of the firm's management.[3] The family by now had a long history of speculation in the communications industry, and that experience reaped rich dividends when they profited handsomely by selling off Associated Communications Corporation to SBC Communications for about $700 million in December 1994. Of that company, they retained control of a small subsidiary (spin-off) called Associated Group. One Associated Group unit, Microwave Services, had collected, for a pittance, several licenses for unused wireless spectrum in the 18-MHz band, and the Berkmans were looking to profit from those licenses. (Later they had to swap the 18-MHz for 24-MHz licenses.)

Radio waves form a spectrum of different frequencies. Many thousands of frequencies comprise the radio spectrum that is used for services like FM radio, short-band radio, microwave communications, and cell phone communications. Usually, the U.S. government and the U.S. Federal Communications Commission decide who gets to use what frequency. For instance, in New York City, the 100-megahertz band serves the Z-100 FM radio system. While typical radio waves like those received by an FM receiver tend to move in all directions, fixed wireless's radio waves go from one point to the second, in a straight line, through small dish antennae.

In the late 1990s, many viewed fixed wireless as a technology that could do the end run around the Baby Bells who owned the "last mile" connections. Last mile connection is the link that connects a central office to a consumer's home or a business premise, and Baby Bells like Verizon and SBC Communications have a near monopoly over this business. The advantage to companies like Teligent was that they would not have to dig up streets and build fiber or copper connections to the customer premises, since hypothetically they could deliver phone and broadband services wirelessly.

Prior to the Telecommunications Act of 1996, the Baby Bell monopoly was much like that of the U.S. Postal Service. Just like with the postal system, wherein only the mailman is allowed to deliver the mail, the Baby Bells were largely the only ones that could provide phone service to consumers. Teligent and other fixed wireless technology companies believed that they could be more like a Federal Express and do it better and cheaper. As fixed wireless didn't need any wires, it was believed that building a vast, spiderweb-like network of rooftop antennae was an easier and cheaper way to provide phone, video, and data, or broadband services.

In the late 1980s, when the Berkmans started accumulating fixed-wireless spectrum licenses, the FCC didn't see much value in them as no one had really figured out a use for them. But from their experiences in the past, the Berkmans knew that any kind of spectrum has value, because there is only a certain amount of usable frequency available for wireless communications, which could be lucrative. It was a purely speculative move.

In 1990 the Berkmans happened to meet Washington, D.C.–based Rajendra and Neera Singh, telecom financiers who had become rich after starting a radio frequency engineering firm called LCC International. The Singhs had just started acquiring licenses on the cheap, under the name Digital Services Corporation. The two families co-invested in a Mexican cellular phone franchise. The contacts would only grow stronger. In the early days it was not obvious what the licenses would be used for, but since it cost next to nothing, the Singhs also became spectrum speculators. In 1993, Congress would pass the mandate that all spectrum needs to be paid for, but by then the two were well on their way.

Rajendra Singh Lunayach, an entrepreneur turned wireless speculator cum venture capitalist, was born in the desert state of Rajasthan in India. He came from a poor family and grew up in a house with no electricity and no phones. His village, Kairoo, is not even on the map. But he made his way to the United States and attended college at the University of Maine. While on a break from college in 1977, he went back to India, where he met his wife-to-be, Neera, while visiting a friend at the Indian

Institute of Technology in Kanpur. They would be married four years later. In 1979, Singh transferred to Southern Methodist University in Dallas, where he learned about wireless communications technologies. In 1980, after getting his doctorate, he went to teach at Kansas State University, where Neera joined him.[4]

At Kansas State, Singh and Neera (also his business partner) started a wireless consulting company in 1983 with a $1,000 investment. They called it LCC Inc., or Lunayach Communications Consultants. He helped design the cellular antenna grids for the United States' first-generation cell phone networks for several operators. LCC would tell cellular companies where to put their antennae for the maximum amount of coverage and best utilization, and they became supersuccessful, gaining customers like McCaw Cellular and AirTouch, two of the early cellular ventures.[5]

Through the 1980s, Singh became an expert on wireless technologies and started to spread his tentacles. Using the profits from LCC International, he started a telecom investment fund, Telecom Ventures, and invested in many start-ups. In 1999, he made it to the Forbes list of richest Americans, coming in at 223 with a net worth around $1.1 billion. "I wanted to come to this country not so much to make a lot of money, but because I could go to school here and become a professor. And then of course, once you come here, then all the cultural influence is to make money," Singh said of his success.[6] Singh was known to push his start-up investments to get liquid quick, so he could get his money and run.

Singh and the Berkmans, who had co-invested in a cellular phone company in 1990, decided to pool their resources in March 1996. The Berkmans and the Singhs were an unlikely team—the former, an old-world rich family, and the latter a member of the nouveau riche. But they were both driven by the dollar.

The Berkmans' investment in Teligent totaled about $50 million. Over the next few months, they looked for the patsy who would turn their worthless licenses into a profitable (at least for them) enterprise. Enter Alex Mandl. "Alex was our number one draft pick," said Bill Berkman.[7] Mandl's compensation package certainly was like that of an NFL draft pick: a $15 million signing bonus, a $500,000 salary, and between 6 and 10 percent of Teligent in stock, worth about $500 million.[8] Like a true believer, Mandl invested a portion of his signing bonus—reportedly

$5 million—in the company. It was the worst investment decision he would ever make.

At the time, many thought Teligent had an especially good shot at taking on the Baby Bells, partly because its unique wireless technology bypassed their wireline networks. Being from an old-school telephone company, Mandl realized the importance of the last mile connection. AT&T had been at the mercy of the Baby Bells who controlled access to the consumer, and Mandl knew the profits that the Bells were milking out of their last mile connections. If a radio technology could deliver the same voice service and high-speed Internet access, Teligent could become a real player in the telecom business. It must also have been reassuring that radio wireless technology had been around for decades and could possibly work in doing an end run around the Baby Bells.

Marconi's radio was a wireless technology and had led to the development of modern communications systems. In most countries outside of the United States, microwave is the key mode of transporting phone calls and video images. As recently as the 1980s, MCI had used microwave technology to go up against AT&T. Mandl must have believed that Teligent's technology could help him build the next AT&T.

Teligent executives wanted to use fixed wireless technologies (i.e., radio frequencies) and become the invisible broadband pipe to the customer's home and business. It was a great idea in theory, but the problem was that whatever the engineers did, the fixed wireless technology simply didn't work because of interference between an antenna and the consumer premise. Window washing gear, people on roofs, flags and flagpoles, or anything that could have blocked the line of sight or could have moved the antenna more than a degree off the main path could render the system useless.

Despite the challenges, Mandl earnestly began putting together a topnotch team. He hired Kirby Pickle, a former MFS executive, to be his chief operating officer, and Keith Kaczmarek, a young wireless Turk who had worked technology wonders for AirTouch and PrimeCo. "It was really an opportunity to be with a rock star team," said Kaczmarek. "Alex brought the credibility and was a strong motivator. He had great vision and was trying to put together an all-star team."

Mandl may have wanted to build the next AT&T, but he never really had a chance to do that. From the day he got to Teligent, insiders hinted that he was under pressure from the Berkmans and the Singhs to take the company public. Mandl wanted to use private capital to grow the company, but Teligent's primary backers wanted to cash out as soon as possible. The all-star team was going to be their ticket to a new fortune!

A company of lesser repute would have been accused of pumping and dumping—that is, promoters talking up the stock and then selling their own stake in the (near worthless) company. But since this was AT&T's crown prince with respectable backers, nobody said or, for that matter, believed that was what was happening.

Under pressure from his board, Mandl reluctantly agreed to a public offering that proposed to raise $125 million in common stock and later about $400 million in debt.[9] In its first S-1 filing with the Securities and Exchange Commission, the company noted that it would start deploying equipment in Dallas, Los Angeles, and Washington, D.C., during the fourth quarter of 1997—around the time the company went public.[10] But Teligent had not even bought any equipment; it was simply testing equipment from Nortel, Lucent Technologies, P-Com, Netro Corporation, and Broadband Networks.[11] It was like putting the cart before the horse, but apparently no one was paying any attention to the fact that this was nothing but a shell company with some licenses, a marquee CEO, and well-known executives.

Before the company went public, Mandl convinced Japan's largest phone company, Nippon Telegraph and Telephone Corporation (NTT), to make a $100 million investment in Teligent. In the summer of 1997, NTT was lusting to get into the U.S. market and was willing to pay any price. For Teligent, this was a great endorsement. Predictably, financial analysts also endorsed the Teligent business plan and deemed the public offering "hot." "It's a sound business plan. They have all the key ingredients: a great set of assets, capital, and an excellent management team. They just have to execute," said Jack Regan, an analyst at Legg Mason Wood Walker in Baltimore, when asked about the prospects of the Teligent initial public offering.[12]

The company's November 1997 IPO, underwritten by Merrill Lynch, Salomon Smith Barney, Goldman Sachs, and Bear Stearns, was a huge success. The stock was offered at $27 a share and surged 20 percent on its first day of trading, giving Teligent a market capitalization of $1.3 billion. At the time, Teligent had about $6 million in cash, about $3 million in revenue, and $79 million in losses. It was basically worthless. "The company had no network at the time—the company simply cashed out on Alex's name," laments a former insider at the company. In his defense, Mandl later said, "It was what the capital environment wanted at the time. Being small wasn't the thing to do back then. We would have had no traction in a very competitive market. People would not have given us a chance."[13]

According to documents filed with the Securities and Exchange Commission, Teligent's equity value and Mandl's payoff were tied. Mandl's contract had clear milestones to boost the stock price. He had to oversee the company to where its stock market valuation would grow to $200 million in August 1997 and eventually to $2.75 billion by 2003. His cut would be 3 percent of the amount by which Teligent's equity value exceeded its benchmarks.[14] In 1999, Teligent was worth $3.2 billion and thus Mandl's worth on paper was $500 million.

Mandl became the newest inductee to the broadband millionaires club. At the time of the offering, Mandl's stake on paper was worth a lot more than that of his longtime nemesis, Joe Nacchio, at Nacchio's new company, Qwest. When Nacchio reported to Mandl at AT&T, it is said that Mandl loathed Nacchio's arrogant style and pompous self-promotion. The two men took subtle swipes at each other at a panel George Gilder's Telecosm organized in Lake Tahoe in 1998, after they had both left AT&T. At the Telecosm panel, Nacchio blasted his former bosses, who included Mandl: "If you're going to step out boldly, you better have the senior guy believe in it. There were a lot of good ideas, but they couldn't get corporate traction."[15] It was quite clear that he was taking a swipe at Mandl, who in Nacchio's mind was part of AT&T's inner circle, which failed to take charge. Tom Evslin, who worked with both Nacchio and Mandl at AT&T, recalls, "Joe worked for Alex, and many people felt that Joe should be the next guy to be the top guy, and that is why the two had a very uneasy relationship."

Blowing in the Wind

Teligent's rocket rise to the top was the perfect launching pad for a fixed-wireless broadband bubble. Within days of Teligent's initial public offerings, there was a substantial uptick in the number of companies that raised fresh capital, either to develop new equipment for the fixed wireless market or to expand their existing services.

One such company was New York–based Winstar Communications, which had its roots in a two-bit merchant bank, Winstar, whose investments also included a small chain of beauty goods and a skiwear maker. In 1994, Winstar's focus shifted to telecommunications,[16] and it started reselling local and long distance phone service. If AT&T was the Macy's of the telecom world, then Winstar was the 99-cent store. Winstar, the merchant bank, was run by William J. Rouhana, an entertainment lawyer turned banker. The bank had a sketchy record[17]—its investment in New York–based American International Petroleum Corporation did well, but its investment in Management Company Entertainment Group, a movie producer and distributor, was a wash.

According to *Bloomberg News*,[18] Rouhana, the man behind Winstar, invested $3 million for an 80 percent stake in Robern Apparel, a New York–based skiwear maker, and became chairman of Robern in February 1991. Three years after it was started, in April 1991, Robern went public at $2⅝ a share. Then in October 1993, Robern changed its name to Winstar Communications and got a new ticker symbol, WCII.[19] The huge moneymaking potential of the yet-to-be-passed Telecommunications Act that would deregulate the industry had caught the eye of Winstar executives. And it was all thanks to Leo George, founder of Avant-Garde Telecommunications. A telecom legend, George, was a 20-year veteran of the telecom business, who had accumulated 30 microwave radio licenses. (Incidentally, he was the same attorney who had fought on behalf of MCI in its battle with the FCC and AT&T.) The spectrum was in the ultrahigh frequency—38 GHz. George apparently wanted to use the technology to link cellular phone base stations, but when Rouhana came across the business plan, he found that there were other things he could do with this technology and the licenses. "Two lines in the business plan haunted me the whole night. They stated that the 38-GHz frequency band is the functional equivalent of fiber, and would be cheaper and easier to install. So—why couldn't I use this to do what I wanted to do?" he later told *Wired*.[20]

These licenses represented the largest block of licenses granted by the FCC in this frequency, and included cities like New York, Los Angeles, Chicago, San Francisco, Philadelphia, Detroit, Boston, Washington,

D.C., Dallas, Houston, Miami, and Atlanta. Avant-Garde could offer services in these heavily populated commercial hubs. In February 1994, Winstar acquired a 16 percent stake in Washington, D.C.–based Avant-Garde Telecommunications for a mere $1.6 million, a bargain basement price, it was later learned.[21] Rouhana would later take control of Avant-Garde completely. With the buzz around telecommunications increasing, Winstar shares surged in 1994 and by the end of the year had nearly doubled in 12 months to about $9 a share. The company's losses nearly doubled as well, from $4.7 million to $8.2 million.

Even this early in the company's life, there was speculation about Rouhana's real intentions.[22] In a news report published on January 4, 1995, *Bloomberg News* wrote: "Under the leadership of William J. Rouhana Jr., chairman and chief executive, Winstar's cash has been used to enrich him and his associates through a series of less-than-arm's-length transactions."[23] *Bloomberg News* suggested there was something improper in the way Winstar's internal divisions handled their finances. Rouhana and his cohorts had accumulated enough options that would allow this coterie to acquire 7.84 million shares, or 43 percent, of Winstar Communications at below-market prices.

Winstar also raised about $225 million from the markets in October 1995 with the help of investment bank Morgan Stanley.[24] It acquired more licenses, issued more press releases, and by 1997 had become a major player in the telecommunications industry.

The next two years would put Teligent and Winstar on a collision course. The two companies would compete for the same customers, using similar hype tactics, and would meet a similar end. Insiders at both companies would make a ton of money, and the investors would be taken to the cleaners. While Teligent's Mandl was telecom royalty, Rouhana clearly was the outsider with a murky past. Mandl played golf with McCaw, and Rouhana socialized with Martin Meyerson, head of M. H. Meyerson & Company, a Jersey City, New Jersey–based investment bank once charged by the SEC for manipulating the stock of Micro Therapeutics Inc.

Sure, Mandl had his pedigree, but Rouhana's crew had the ultimate weapon: Jack Grubman, the Salomon Smith Barney analyst, who pumped up Winstar to new heights before both the company and the analyst fell on hard times. On January 9, 1998, Grubman formally started covering Winstar with a buy rating and issued a $71-a-share price target. At the time, Winstar was trading at $19.50 a share, but Grubman expected the stock to be worth $71 by January 1999. For Rouhana, sucking up to Grubman was well worth it, since Grubman could open doors for Rouhana's dinky outfit. Grubman, in turn, had a vested interest in helping Rouhana because his loyalty could bring big banking fees to Grubman's employer. With the increasing importance of the wireless market, Grubman needed a close relationship with a wireless company to secure his position as a top telecom analyst. Teligent, thanks to its blue-blood heritage, didn't really need Grubman. Winstar fit the bill.

Altogether, Salomon Brothers helped Winstar raise $5.6 billion, in the process earning more than $50 million in fees.[25] Rouhana's defining moment came in October 1998, when Lucent Technologies announced that it would provide $2 billion in equipment and financing to the company. Rouhana had worked out a deal with Carly Fiorina, a hard-charging saleswoman who was known to cut corners in order to meet her sales quotas. (Of course, she left while the going was still good and ended up running Hewlett-Packard.) Rouhana was giddy, boastful, and tripping over himself—Winstar stock was at $16 a share (split-adjusted), and the media was hanging on to every word this self-described technology visionary had to say.

He boasted, "This is a defining moment for Winstar. Lucent's major commitment of expertise and financing, combined with the overwhelming speed-to-market and cost advantages of Winstar's business model, clearly propels us to the top of the competitive local exchange carrier industry. With Lucent's network knowledge behind us, we are positioned to be the first competitive carrier to create a nearly ubiquitous end-to-end broadband network in the top 100 world markets."[26]

But what Rouhana didn't mention, and what no one bothered to question, was that this fixed wireless company was nothing but a reseller of traditional wired phone services, which it was buying from regional Bells and other established players. Like an Amway salesman, it was buying wholesale and selling retail. Winstar was really an old-fashioned telephone company, not the new economy, next-generation broadband wireless firm it claimed to be. According to an SEC document filed at the time of the Lucent announcement, about 18 percent of its lines were wireless—the rest were traditional phone and data lines. Its competitor

Teligent was no different—it was leasing T-1 lines from regional phone companies by the thousands, paying upwards of $300 per T-1 line per month, wholesale.

After the Lucent deal, Winstar stock took off and eventually peaked at $60 a share, giving the company a market capitalization of over $2.5 billion, and Rouhana became the new media darling. Like Gary Winnick, Rouhana had found a good thing and turned himself into a visionary who knew all the answers to the problems facing broadband providers—except those of Winstar, whose losses were mounting. The company lost $700 million in 1999 compared with $488 million in 1998; the company's debt increased to $3 billion from $2 billion.[27]

But Rouhana and his sidekick, Winstar chief operating officer Nathan Kantor, were rich. Along with other Winstar executives, the two had cut themselves nice little deals. In 1999, Rouhana made $537,002 in salary and $858,587 in bonus.[28] The bonus was mostly for doing his job, for which he was already getting a considerable salary. But the annual report said he was awarded a bonus for securing "significant additional funding by entering into an agreement for a $900 million private equity financing"[29] and for help with the "significant appreciation in enterprise value and common stock market price from the prior year."

Wireless and Free to Fly

The success of Teligent and Winstar on the financial markets gave a certain legitimacy to fixed wireless broadband technology. In 1999, both Sprint and WorldCom got into that business, and reportedly made overtures to buy Teligent. Other lesser-known copycats also emerged, like Advanced Radio Telecom (ART), which attracted $251 million in investment from Joe Nacchio's Denver-based Qwest Communications and from nine other venture capital firms. Cisco Systems kicked in another $175 million for buying equipment such as routers and switches for ART's network. Even Nextlink, a company started by Craig McCaw, dabbled in fixed wireless technology.

This, in turn, created a bubble in the fixed wireless equipment makers' sector, which included companies like Netro and P-Com, whose gear, companies like Teligent and Winstar would presumably be buying to establish their networks. Thanks to its contract to supply equipment to

Teligent, Nortel spent almost $416 million to buy Canadian equipment maker Broadband Networks. "We see a huge market for fixed wireless access and are gearing up accordingly," said Guy Gill, vice president and general manager of access networks at Nortel.[30] Companies like P-Com and Netro, which had been early players in this space, caught the attention of media and stock market speculators. Netro raised almost $40 million in its initial public offering in August 1999.

Rising tides lift all wrecks, and soon fixed wireless equipment providers—Triton Network Systems, Airspan Networks, Giganet, Floware Wireless Systems, Vyvo, BreezeCOM, and Adaptive Broadband—were tapping the markets. It also helped that market research firms were predicting that globally, fixed wireless broadband equipment sales would increase from $442 million in 1999 to $5.38 billion in 2004, and that service revenues would rocket from $740 million to $16.2 billion in the same period.

Mandl Got the IPO Blues

Despite a successful IPO, 1998 was a tough year for Teligent. Even before the company went public, it had to swap its 18-gigahertz licenses for 24-gigahertz licenses, on the orders of the Federal Communications Commission. Apparently McCaw and Bill Gates, who were funding Teledesic, a low-orbit satellite broadband company, complained to the FCC that their networks would face trouble because of the 18-GHz licenses Teligent had.[31] The FCC, after a lot of bickering, had to come up with a compromise and swap the frequencies. The swap, however, resulted in the Singhs and Berkmans getting four times the frequencies they originally owned. The move may have given Teligent more spectrum, but it also created all sorts of technical problems for them.

At about the same time, Teligent also made a bad technology bet. They wanted to use point-to-multipoint fixed technology, a key differentiator from main rival Winstar, which used traditional point-to-point, fixed wireless technology. Point-to-multipoint technology was said to be cost-efficient and could offer more flexible bandwidth. Point-to-point equipment is much like a walkie-talkie that works when two people are standing in a straight line. But multipoint fixed wireless technology was like a radio set, which sends out many radio beams at once and also receives multiple beams at the same time.

The technology was not ready for prime time, and many times it simply did not work. And if it somehow managed to work, the high-speed access for point-to-multipoint needed so much radio-beam power that none of the available solutions could achieve the phone-company quality that Teligent was promising its customers. Nortel, desperate for revenue, promised to deliver equipment that would make Teligent's network a reality. In order to win the contract, Nortel kicked in about $780 million of vendor financing, a kind of loan to Teligent to buy equipment from Nortel.[32] A former executive who was closely involved with the network buildout confessed that the problem was that "we put too much on stake on this technology. We spent too much time on this technology, trying to make it work. Some of the fundamental challenges were around technology. Nortel had problems delivering the right product." In fact, its technology troubles were Teligent's worst-kept secret, and it is surprising that not one Wall Street or industry analyst picked up on it.

By September 1998, Teligent had lost a whopping $328 million on revenues of $4.3 million—and the network still didn't work. A former Teligent engineering employee said, "It was way too complex to install. It was taking us 90 days to set up service and that was way too long."[33] "What I could not figure out was why we were spending $30,000 per site for a 3-cents-a-minute phone service," a former executive said.

Insiders say the Teligent management was asleep at the wheel and, instead of trying to fix its technology problems, was busy signing up leases with buildings where it would provide access and install its antennas, which are the size of a dinner plate. In 1998, the company spent millions on equipment and leases. Marketing campaigns and other costs were zooming up higher. With over 2,500 employees, Teligent spent almost $122.3 million on salaries and other compensation. "We were bleeding money at almost $100 million a month," a former insider said.

The company had rolled out an aggressive marketing campaign, and Teligent's black-and-red ads were everywhere—at Yankee Stadium in New York, at FedEx Field in Washington, and even in Times Square. By 1999, Teligent was one of the best-known companies in America—with nothing to sell. The company had spent billions of dollars in building operations in 42 markets across the United States and Europe. But the demand was just not there, and almost everyone was fighting for the same customers. Either Teligent didn't know that, or it deliberately overlooked that vital fact.

Still, Teligent stock kept climbing, helped by the liberal praise of Wall Street analysts, including Grubman, who was going gaga over fixed wireless companies. The hype machine at Teligent was already in high gear. The company was pumping out press materials faster than it could pump out those bits and bytes. Mandl and his cohorts were busy talking up the small deals that they did manage to score. The company figured it could get a lot of press mileage out of providing broadband access to football teams like the Washington Redskins, Arizona Cardinals, and Chicago Bears. This was the brainchild of senior management, including COO Kirby "Buddy" Pickle, a devout football fan. In 1999, he reportedly chartered a private jet and flew to the Super Bowl, where he hung out with some of his salespeople, and then took his family on the same jet to Hawaii for a vacation. Teligent also had box seats at other stadiums where the company could entertain important clients. Everyone was impressed.

In the two years since it went public in November 1997, investors bid up the stock to $100 a share and gave Teligent a valuation of $3.2 billion—all this for a company whose technology simply didn't work and which didn't have many customers.

The Blind Lead the Blind

In the initial stages of the telecom boom, customers like the new dot-com start-ups were willing to try out the cool new offerings from fixed wireless companies. The rush to the Internet created a backlog at the Baby Bells, which often took as long as three months to install a single high-speed connection. The dot-coms, which desperately needed high-speed access to the Web, assumed that companies like Teligent and Winstar would provide a better and speedier option.

Hardly! For most of 1998 and 1999, it was taking Teligent and Winstar somewhere between 30 and 90 days to get their customers online. When their own wireless systems didn't work, they just installed T-1 lines, which they bought from none other than . . . the Baby Bells. In 1999 Teligent had leased 30,000 T-1 lines, which meant the pesky 90-day wait just wouldn't go away!

So how could Teligent and Winstar spin their customer statistics for Wall Street at a time when Wall Street was valuing their ascendant stocks based on the number of customers and projected revenue? Not even the most bullish analyst expected the duo to make any money until 2007. So it was Grubman to the rescue with another market-driven, dead-wrong metric that would count building leases as a part of customer growth. How did that supposedly work? Access to the buildings where Winstar and Teligent could sell their services meant more tenants, that's how. So Teligent signed leases and built a network, but it could never sign up enough customers to have meaningful revenue.

Some internal documents show that in April 2001, weeks before it filed for bankruptcy, Teligent was making about $9 million a month from 10,306 buildings it had wired. Only $2.5 million a month was coming from the vaunted wireless technology. Teligent had spent billions on building a network that was producing a few million in sales, not profit. It had access to 10,300 buildings at the time, but only 16 percent of the total tenants in those buildings had signed. Customers paid roughly $800 a month to Teligent, which in turn paid $600 to the local phone company for two T-1 connections, one for voice and one for data, just in case the wireless network didn't work—which, most of the time, it didn't.

Winstar had similar customer problems, but the company tried to solve it with a time-tested strategy: You scratch my back and I'll scratch yours. In December 1998, Tulsa, Oklahoma–based fiber-optic carrier Williams Communications announced it would pay $400 million over four years to use Winstar's wireless broadband network. In turn, Winstar promised to buy fiber-optic capacity worth $644 million from Williams over seven years.[34] So what if none of the little guys—the real customers—were buying the services as had been hoped. Showing revenue growth was fairly easy!

So easy, in fact, that Winstar did it again. In 2000[35] and 2001, Winstar invested $145 million in a business-to-business dot-com called Wam!Net, which included $95 million in cash.[36] Wam!Net was known on the trade-show circuit in the 1990s for giving away pairs of Converse high-tops to people who sat through their presentation at trade shows. A few magazines used their integrated services digital network (ISDN) service to deliver pages to their printers.

Wam!Net, in turn, promised to buy Internet access and other services from Winstar and promptly paid back $20 million to Winstar. It also promised that every three months it would start paying back $5 million and then increase it to $25 million for the following years. How much was the total amount they were going to give back? Well, in the grand scheme of things, no one really cared about the paybacks! In return, Winstar would provide broadband access to the company. Confused by the new-economy math? So were investors, who either never noticed these deals or chose to ignore them.

After all, Winstar's annual reports were showing huge revenue growth—up from $445.6 million in 1999 to $759 million in 2000. The losses had ballooned from $639 million to $870 million. At the end of 2000, Winstar's customers owed the company $226 million. Debt was around $5 billion and the company was spending nearly $342 million on interest payments—but as long as Wall Street was willing to gamble, and Rouhana's gang had King Jack on their side, borrowing money wasn't going to be a problem.

Over at Teligent, Mandl, too, was knee-deep in problems. The financial situation at Teligent was getting more precarious—the company had a mere $31 million in revenue in 1999 and losses of as much as $528 million. A year later, the company had sales of $152 million, but losses of $808 million. It admitted it was on a deathwatch. It needed cash—fast!

To make things worse for Teligent, the stock and debt markets—previously the company's primary source of funding—nose-dived. By March 2000, the dot-com bubble had burst and cheap capital was no longer available from formerly generous investors. The run-up in the broadband sector had been driven by the dot-com bubble. As the dot-coms grew, they needed more bandwidth and new services. As they were mostly small companies, they decided to look to the upstarts like Teligent and Winstar. It was a brave new world where the little guys ruled—or thought they did.

Brother, Can You Spare a Dime?

Even as Teligent's networks were bleeding money, its top executives and board members were, with scant regard for their investors, living it up. Senior executives got thousands of shares and options, while the actual workers got 500 shares each. Loans and unlimited expense accounts were part of senior management compensation packages. It is said that everyone made money except Mandl. Insiders sold millions worth of stock,

with a few notable sellers: Raj Singh, COO Buddy Pickle, and human resources chief Steven Bell. At one point in Teligent time, Pickle was worth
$150 million on paper. The story goes that when Buddy and Bell were
selling shares, Teligent employees were told it was so the company could
give them (the employees) more shares.

"Teligent was built for personal gain, and some of the executives
robbed the company blind," said a former engineer with the company.
There was something called the President's Circle Club at Teligent,
whose members had unlimited and unaudited expense accounts, private
jets, and other extravagant perks. For instance, the President's Circle
Club met in Hawaii in 2000, and most people stayed at the Grand
Wylie, a luxury hotel. Pickle got a $10,000-a-night room! Pickle, insiders
say, wasn't terribly effective, as he had the attention span of a 10-year-old,
according to some. Others say he was charming, well dressed, and funny.
Alex was out of the loop because he was busy selling the vision and trying
to raise money. "Alex simply trusted people too much and he really did
not have much control," said one insider. "Alex is such a chivalrous guy
that he gave it to Buddy. But Buddy was busy playing golf and hanging
out with his friends from MFS."

The company would go through hierarchical changes every six
months—later, even more frequently. Many executives quit, disgusted.
"Buddy is the most charismatic guy, a great motivator, but he wasn't doing anything for the company, all for himself," recalled a former insider.
"There was no control and no one was watching the store." A perfect example of corporate waste was the equipment the company bought. At its
peak, the company employed about 3,400 people but had about 7,000
laptops, mostly manufactured by Dell Computer. (Mandl happened to
be on the board of Dell Computer.)

Many believe it was bad technology and poor management that became Teligent's undoing. Others say the board did its bit to accelerate
Teligent's demise. In 1999, when WorldCom and Sprint both made
overtures to buy Teligent, nothing happened. Mandl had a chance to sell
the company and get out when the going was good, but he didn't pull the
trigger. He should have, for hell was about to break loose inside Teligent.
"Teligent was hampered almost from the start by internal strife. Mandl's
ambition put him in conflict with Teligent's board, which was controlled
by the company's founding family," wrote *Fortune*.[37]

Mandl realized that he was backed into a corner and that Teligent needed more cheap cash. So he used his old-economy connections. In November 1999, he managed to convince the buyout funds Hicks, Muse, Tate, & Furst, and other institutions like Chase Capital Partners, Deutsche Bank Capital Partners, and Olympus Partners, to invest a whopping $1 billion in the company. Hicks Muse, which had already invested in another local phone company wannabe, ICG Communications, and in DSL provider Rhythms NetConnections, was taking the lead in this round of fresh investment.

The only problem was that Teligent's board wasn't willing to accept this new investment. It isn't hard to see why: If Teligent took the money, Associated Group's 40 percent stake in Teligent would be diluted, which would have a negative impact on a deal the company was creating with Liberty Media. The Associated Group had in June 1999 decided to sell Associated to John Malone's Liberty Media Group for $3 billion; the deal was slated to close in January 2000.[38] Malone, who had already sold TCI to AT&T for $48 billion in 1999, was busy putting together a new content and communications–focused collection of assets under the umbrella of the Liberty Group. With Associated valued at $3 billion, the Berkmans' Teligent assets would be worth $1.4 billion.

Mandl didn't have a chance against the machinations of the board members and fellow executives.[39] There were too many big egos in the company. Despite being fairly successful people with country-club lifestyles, they would squabble over petty issues like designated parking spots. It was also rumored that there were conflicts in the board, as the Singhs and the Berkmans pursued different agendas. Insiders have maintained that while Raj Singh was pushing for aggressive growth, the Berkmans decided to cash out while they could. Singh, of course, had enough time to slowly sell a big chunk of his share in the company.

This was the beginning of the end.

"The Berkmans had the right to make any judgment they wanted to make about the funding deal. In hindsight, it was a terrible decision to turn it down (the investment from the buyout funds) because Teligent would have been fully funded, but at the time, some good arguments undoubtedly could have been made," Mandl later commented.[40] Instead, the company took only $500 million, half of the planned amount from the proposed investors, but it wasn't enough. Teligent chief financial officer Abe Morris quit, perhaps being the first to know that Teligent was sinking faster than the *Titanic*. Ironically, as this was going on, the company's stock market valuation was skyrocketing—in April 2000 it was valued at $6 billion. Mandl had been pleading with the board to raise more money, and even wanted to have a $250 million secondary stock offering, but the board wasn't interested and resorted to dragging its feet. The proposal was ultimately shot down.

Winstar, on the other hand, had no problem finding well-heeled investors to raise money from. It had a $2 billion vendor financing agreement with Lucent from 1998 and, in December 1999, raised $900 million in total from Microsoft; Credit Suisse First Boston's private equity group; Welsh, Carson, Anderson & Stowe, an investment firm; and Cascade Investments. In March 2000, with its stock nearing an all-time high of $65 a share, Winstar retired its old debt with new junk-bond financing worth $1.6 billion. Salomon Smith Barney and Credit Suisse First Boston (CSFB) led the offering, and some believe that, thanks to Jack Grubman, Salomon walked away with millions in fees.

Even as the going got tough toward the end of 2000, Winstar raised another $270 million by selling preferred stock to Microsoft, CSFB, Welsh, Carson, and Compaq Computer. Compaq and Cisco Systems also agreed to finance equipment for Winstar. Siemens Financial, a division of a German telecom company, lent Rouhana's crew another $200 million. The quid pro quo was that Winstar would buy $150 million in equipment and services from the Germans. With each deal, Winstar put out a strutting press release. (See Figure 6.1.)

The high-flying Winstar stock was, however, losing steam and many other new carriers had started to go bankrupt. Investors were carefully scrutinizing the financials of phone companies. By December 2000, Winstar stock fell to $12.50 a share. "The stock prices of the broadband carriers, such as Teligent, Nextlink and Winstar, have all taken a fall, some of a severe nature. No longer are they the darlings among investors," wrote *Broadband Week*, a trade publication, late in 2000.[41] Winstar, in the previous five years, had done two things—lost $2.3 billion in order to bring in $1.6 billion in revenue, and made some insiders very rich. The big losers in the Winstar saga were, of course, the investors who had trusted their pension funds and 401(k) dollars to the so-called professionals at mutual

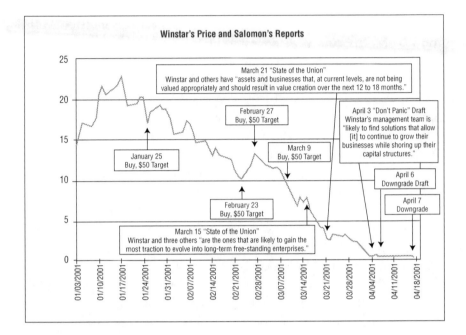

Figure 6.1 Grubman's Finest

Jack Grubman kept the buy rating on Winstar till the very end.

Source: NASD Department of Enforcement v. *Jack Grubman and Christine Gochuico*, Disciplinary Proceeding number CAF020042

fund companies like Janus (which put nearly $200 million in Winstar), Putnam Investments, MFS, and Fidelity, which, along with scores of other funds and pension trusts, bought Winstar debt.

Winstar's supporters were busy putting the smiley emoticon on e-mails that relayed bad news. Grubman tried his best to keep the stock flying high by writing positive reports about the company that he himself didn't believe in. In an e-mail exchange with an institutional investor on February 12, 2001, Grubman's associate, Christine Gochuico, wrote, "I know where you stand, but it might be time to cover your short! ☺ Seriously, I hope you make a lot of money on this one but I also hope it doesn't go to zero since we've been so vocal on it."[42] But no amount of smileys could save Winstar.

Three Strikes and You're Out

By April 2000, Teligent was sinking fast. Desperate for an exit strategy, late-stage financiers were scrambling to come up with a strategy. Hicks Muse proposed a three-way merger between ICG, Rhythms, and Teligent—its three telecom investments that were swirling down the toilet. It was a stupid idea, since ICG was on the verge of bankruptcy. Still, ICG and Teligent decided to swap $62 million of stock. (Later, Liberty's Malone, who by then owned Associated Group's Teligent stake, figured he could restructure ICG on his own, but it didn't work. It was like running a fool's errand. ICG went belly-up.)

Meanwhile, high-flying Buddy Pickle had had enough—he quit with his millions in the bank. Like a martyr, Mandl decided to give it one more shot, and in December 2000 managed to put together a deal: $700 million in new funding from current investors like Liberty, Hicks, Muse, and the Singh family. Lucent Technologies decided to kick in $200 million for equipment and, as a back-up plan, Mandl lined up another $250 million from the Rose Glen Capital Group, a Bala Cynwyd, Pennsylvania–based vulture investment group that specialized in investing in troubled companies. The basis of the Rose Glen deal was that Teligent stock would have to stay above $2 a share before Rose Glen could show Mandl any money. It was a losing battle for Mandl.

In March 2001, Mandl walked out of Teligent's offices.

By April 2001, Malone, having witnessed his telecom holdings evaporate, sold his Teligent stake, once worth $1.4 billion, to IDT Telecom, a little known start-up, for $37 million. IDT was owned by Howard Jonas, a former hot dog vendor turned telecom entrepreneur from the Bronx, New York. Jonas also bought Hicks Muse's Teligent holdings in exchange for IDT stock, according to *Fortune* magazine.[43]

In May 2001, Teligent declared bankruptcy, with $1.4 billion in long-term debt and $1.2 billion in assets. A former engineer was amazed to hear that the company had that many assets. A few days later, Advanced Radio Telecom and Winstar went bankrupt as well. Jonas offered $38 million for Winstar's assets. He figured he could build a fixed wireless empire on the cheap—after all, he wasn't burdened with debt like the defunct operators he sought to acquire.

So why did Teligent really fail? Insiders believe there were at least five reasons that in tandem led to the demise of the company:

1. Taking on the Baby Bells was underestimating formidable competition.
2. Poor management and corporate waste.
3. Too much capital made Teligent sloppy.
4. Terrible technology bets.
5. A hostile, greedy board that only wanted to get rich, the company be damned.

At the end of the five-year Teligent saga, the Singhs, the Berkmans, and a handful of executives had millions of dollars more in their banks. Meanwhile, the losers included Hicks Muse, NTT Corporation, investment banks like Chase (now JPMorganChase), and a lot of little investors—and Mandl. He should have paid heed to the words of Antiphanes, an ancient Greek dramatist who said, "The quest for riches darkens the sense of right and wrong."

7

NOBODY@HOME

It was a drama worthy of the deft touch of Aaron Spelling, the showman behind shows such as *Beverly Hills 90210*, *Dynasty*, and *Melrose Place*. A scion of an American publishing family, a world famous financier, a NASA scientist, a Machiavellian cable cowboy, and a chief executive who looked like he had just stepped off the set of *Hawaii 5-O* were the stars of a business thriller like none before. Ego, greed, boardroom battles, wild parties, and billions of dollars—it was all part of the life and times of a start-up that coined the word *broadband*.

It had everything going for it—exclusive agreements with 21 cable companies, including the nation's second largest cable operator, TCI, along with billions of dollars in capital and thousands of consumers who were dying to get onto its network. Welcome to Excite@Home, the first and perhaps the single most defining broadband company in the world.

In a brief six years, it went from an idea on a paper placemat to a company worth $35 billion, and then a nightmare. The company's now-empty headquarters next to Highway 101, the main artery for Silicon Valley, is a constant reminder to commuters of the hubris that personified the broadband bubble. More than anything, the empty building in Redwood City is also a symbol of a culture clash between Silicon Valley and the rest of the world. It was one of the most spectacular failures in the broadband sweepstakes.

■ ■ ■

Charles Moldow could hear the dastardly bell again. Something needed to be done—it was ringing too often, the 31-year-old former cable company executive thought to himself, and then smiled, wistfully thinking about September 1996, when the bell would ring, at best, three times a day. Every time the bell rang, it signified that @Home Network (as it was then known) had signed up yet another customer for its high-speed Internet service.

Running over the coaxial cables owned by cable companies such as Tele-Communications Inc., Cox Communications, and Comcast, the @Home service was the fastest way to download music, watch videos, or get on the Internet in the spring of 1997—the company's connections were 700 times faster than the 14.4 kilobits-per-second connection one got using a phone line and a modem. Moldow, like many of his 200-odd colleagues, marveled at how quickly the one-and-a-half-year-old @Home had grown.

Excite@Home was a pioneer in offering high-speed Internet access to consumers in the world. In 1995, when John Doerr, a renowned venture capitalist and general partner with the Silicon Valley venture capital firm Kleiner Perkins Caufield & Byers, first met Bruce Ravenel, a senior vice president with Tele-Communications Inc. (TCI), Excite@Home was merely an idea.

TCI was one of the largest cable companies in the world. It was run by maverick John Malone, who had cobbled together many disparate cable systems to build a network that could reach 33 million homes in America.[1] TCI was a fearsome company that had always aspired to become a leader in interactive television. In the early 1990s, when the information superhighway and interactive television were the talk of the town, TCI had tried to merge with Bell Atlantic, but the U.S. Federal Communications Commission scuttled the merger.

No wonder Doerr found TCI executives receptive when he suggested Internet access over cable lines using a new type of cable modem. It was a revolutionary idea at the time, for no one had any idea about the potential of the Internet as a consumer phenomenon. Doerr had been an early investor (via his venture fund) in a company called Netscape Communications, which had developed a Web browser for consumers. He had also zeroed in on the problem of high-speed consumer access to the Internet.

Doerr, whose firm had also backed the Internet service provider America Online (AOL), knew that dial-up technologies of the kind AOL offered would only go so far because they were inefficient. They tied up consumers' phone lines, and downloading pages through them was slow and tedious. Doerr's solution was "broadband," or a big fat pipe that could download information from the Internet at blazingly fast speeds.

A former Intel executive, Doerr was well known inside Silicon Valley as the man whose insane ideas turned into multibillion dollar opportunities. He had funded Genentech, a biotechnology company; he had backed Sun Microsystems when no one thought there was money to be made in UNIX-based workstations. So when he talked, people listened, lest they miss out on the next big thing. One of those who paid attention was John Malone.

Doerr traveled to Denver, Colorado, to pitch his cable Internet access idea to Malone. Doerr wanted Malone to convince his peers in the cable business—Continental, Cox Communications, and Comcast, also known as the three Cs—to come on board as partners in this still-unnamed venture. He had picked the right guy. It was known that in the cable business there is no better arm twister than John Malone, who ruled the cable universe like a ruthless dictator. TCI and the three Cs were powerful forces, and they were all looking to boost revenue that had become stagnant because of the cable market's saturation. Plus they needed something new to sell to customers who were being lured by the more nascent digital satellite broadcast devices.

If Malone is the arm twister, Doerr is the man with the silver tongue. His pitches are famously hard to resist. As one wag quipped, his goals are packaged like a Christmas gift wrapped in velvet and tied with a silk bow. "As a salesman, he's so good he can sell you just as easily on bad concepts," said Jim Clark, the chairman of Netscape and the founder of the server maker Silicon Graphics.[2] Doerr can make even the most mundane technology sound like the second coming of the personal computer.

"Doerr brought his particular vision to the table that said, Silicon Valley start-up, independent management, headed toward a public offering, and creating a nationally branded service that the cable industry would deliver," said Ravenel.[3] Naturally, Malone said yes to his idea and took a 75 percent stake in the start-up. In March 1995, @Home Corporation was born. Even the start-up's name was Doerr's idea. In the beginning, the company was based in a ramshackle office in Mountain View, California. It later moved to Palo Alto and then to Redwood City, both Silicon Valley towns.

Now Doerr needed a marquee name to head up this project. Enter William Randolph Hearst III, the scion of the Hearst publishing empire, who had recently quit his job as a publisher of the Hearst Corporation–owned *San Francisco Examiner* and had joined Kleiner Perkins as a general partner. Hearst signed on as the company's temporary chief executive. He was a Harvard graduate, an eloquent promoter of technology, and had a pedigree to match. In short, he was the perfect CEO for @Home.

Rush Limbaugh of the Internet

Doerr's start-up had a name and a CEO, but no network. He needed to find someone to build it, and it had to be someone who was the very best in the business. That's where Milo Medin came in. Medin, a 32-year-old computer engineer and networking genius, was part of the new network elite, one of a few hundred people around the globe who knew how the Internet really worked. At the age of 23, he had helped build NASA Science Internet, which linked researchers in 16 countries across six continents.

Born in Fresno, California, to Serbian immigrants, the young Medin's fascination with computers and, later, networks took him to the University of California in Berkeley, where he studied computer science. Medin, who once called himself a "farm boy from Fresno raised on 20 acres of grapes,"[4] went to work for NASA in Silicon Valley when he was a college senior. Upon graduation, he joined NASA as a full-time employee. By 1988, he was the team leader for a project that proposed to link together different NASA facilities in the Silicon Valley region.

At the time, there was a plethora of networking standards, most of which weren't compatible with each other. They were like the modern-day cell phone networks. For instance, if you switch service from AT&T to Sprint PCS, you need to buy a new phone. It was the same with the networks then. To get around the problems of interoperability, Medin backed a standard called TCP/IP (which stands for Transport Control Protocol/Internet Protocol). TCP/IP was open and not dependent on any operating system—it worked well on UNIX, Windows, or Macintosh operating systems, most networking devices could understand it, and it was easy to build new applications on top of it. "He was simply one of the loudest and most effective voices saying TCP/IP was the only way to go. And if it wasn't for that, there wouldn't even be an Internet," said Steve Wolff, then director of network for the National Science Foundation, of Medin.[5] His support of the TCP/IP standard made him a force to reckon with in the world of networks and the Internet. Some called Medin the Dr. Strangelove of networks, while others dubbed him the Rush Limbaugh of the networking protocol (because of his conservative views on U.S. foreign policy).

Getting Medin to join the new start-up as the network architect would be a coup. Doerr left many phone messages on Medin's voice mail, but Medin ignored them, assuming KPCB was a law firm. But Doerr was relentless in his pursuit. Doerr and Hearst finally got him to meet them for breakfast at The Good Earth, a health-food restaurant in Palo Alto, California.

Over wild rice pancakes, Doerr and Hearst sketched out their vision of the network on a piece of paper. They were clearly not prepared for what was to follow. Medin, a blunt man with a take-no-prisoners approach, told Doerr and Hearst their business plan wouldn't work. In an interview with *Fortune* magazine, Medin recalled that Doerr and Hearst looked as if he "had just run over their puppy with a truck."[6]

Medin told them that linking computers to cable modems to a wide-open network would overwhelm the network and would create outages. He then went on to give them a lesson in Internet architecture. "In two minutes he had changed our business model," Doerr later told *Wired* magazine.[7] "There are only half a dozen people in the world who make the Internet work and Milo is one of them. We absolutely needed him."[8]

Instead of Doerr's plan, Medin proposed building a souped-up network that would run on Internet protocols. He then proposed they buy bandwidth from companies like Sprint, MCI, and UUNet, and then build special areas on the network where content would be stored and cached. (A few years later, companies such as Inktomi and Akamai made millions off this caching concept.) Medin's plan would cost more than $125 million.

Doerr said yes. And the engineer in Medin couldn't resist the fact that

Doerr agreed to such an audacious and challenging proposal. He signed on as the first employee of the company. Kleiner Perkins invested $2.3 million, and TCI weighed in with $7.7 million to get the company started. In 1995, @Home began work on the first consumer broadband Internet access network. There were no standards, hardly any cable modem providers, and no one making equipment to support the kind of network @Home wanted to build.

While the early employees were trying to come up with solutions, Doerr, Hearst, Ravenel, and several others were trying to convince other cable partners to come on board. Time Warner, then the second largest cable company, declined to get involved, as the company had started experimenting with its own cable modem service, called Road Runner, in Elmira, New York. Continental Cablevision, one of the three Cs, decided to go its own way as well. But Cox and Comcast signed on, and each picked up a 14 percent stake in the company.

That meant that @Home now had access to 40 million of 105 million American homes and could theoretically offer them high-speed Internet access. The companies worked out a deal by which the cable companies would charge $35 a month for the service and then split the revenue— @Home would get $15 per customer, and the cable operator would get $20 a month per customer. It seemed like a perfect situation. Everyone was going to make money and @Home had something all start-ups crave: a virtual monopoly over a market.

High-Speed Hype

But things didn't go as well as planned. In what was typical of the bubble era companies, @Home became a classic example of a syndrome of over-promising and underdelivering. In a press release announcing the company, @Home claimed it would roll out its service in early 1996 and sign up a million customers by the end of 1996. It was a statement the company would come to rue, as it quickly became a public relations nightmare for the young start-up.

"When Will Hearst was predicting one million subscribers, I did some back of the envelope calculations and knew right away it was a mistake. In 100 business days, assuming a two person installation team could do four homes per day, we would have needed 50,000 installers working full time," Peter Huber, the company's 99th employee and later vice president of engineering, told the *Industry Standard*, the now-defunct chronicler of all things dot-com.[9] Technology, it seems, wasn't ready for prime time, and it wasn't until September 1996 that the company would launch its service.

Meanwhile, in the background, a new drama was unfolding. The senior management of @Home was paying the price for being in bed with the cable guys, thanks to Bruce Ravenel, who as TCI's representative was on @Home's board of directors. Insiders called him Malone's hatchet man. He went after members of the senior management team, disparaging them one by one. " 'We don't negotiate with you, we own you,' and he (Ravenel) didn't mean @Home, he meant *you*," recalls Sean Doherty, who ran @Work, a division of the company servicing the small- and medium-sized sections of the business. Others who faced Bruce's ire included Dean Gilbert, who was running the @Home business on a day-to-day basis.

Ravenel and the other cable guys plotted to merge @Home with Teleport, a Staten Island, New York–based local phone company, and turn the merged entity into a broadband powerhouse. In 1996, teams were going back and forth between @Home's Silicon Valley offices and Teleport's offices in Staten Island. TCI, Cox, and Comcast all had a stake in Teleport. But the plan didn't go anywhere, and in the end, AT&T bought Teleport for $11.3 billion in 1998. (This was the first time—but not the last—that Malone and Company took Ma Bell's money. TCI ended up with a 10 percent share in AT&T, since it was one of the main backers of Teleport, and got AT&T stock in exchange for its share.)

Meanwhile, Hearst had quit as @Home's CEO, explaining he was a temporary chief executive anyway. It was time to bring in the heavy hitters. Doerr assigned the uber-head hunter David Beirne (now a partner at the Silicon Valley venture capital firm Benchmark Capital) to go find him a chief executive. Among those who interviewed with @Home were guys like Ed Bennett, who ran Prodigy; David Dorman, then CEO of Pacific Bell and currently chief executive of AT&T; and Joe Nacchio, then an AT&T executive, who later took the top job at Denver-based Qwest. (Nacchio didn't want to move to Silicon Valley because his kids were going to school in New Jersey. Qwest had offered to fly Nacchio back to New Jersey from Denver every weekend.)

One man who stood out in Doerr's mind was Thomas Jermoluk, president of Silicon Graphics. But Jermoluk didn't want the job. At the time, Silicon Graphics, which makes huge computers that do everything from crunching weather satellite data to making *Jurassic Park* come to life, was a $3 billion-a-year sales operation. Jermoluk was making $1 million a year and didn't need to go work for some pesky little start-up. But you can never say no to John Doerr, and eventually Jermoluk signed on the dotted line over a meal of Italian takeout in Doerr's office. The deal gave him enough stock options that if all went according to plan, Jermoluk could be worth hundreds of millions of dollars.

TJ@Max

Thomas Jermoluk is known to Silicon Valley insiders as TJ and is one of the more colorful characters to inhabit Silicon Valley. Born in New Jersey, TJ grew up in Hawaii and still has a bit of the islands' laid-back attitude. He is also known for his wild parties, no doubt a carryover from his early years in Hawaii. There were whispers about TJ not exactly being the kind of guy you would want your daughter to bring home. Though married once, TJ is frequently seen in the company of beautiful women, some so young that one wonders if he has ever heard of the old British rule that gentlemen shouldn't date women who are younger than half of their age plus nine. Jermoluk's windswept blonde hair, colorful shirts, and charming, easygoing manner have made him one of the most gossiped-about executives in Silicon Valley. He is like a boy who never grew up. TJ loves hanging out with buddies like Scott McNealy, the toothy CEO of Sun Microsystems.

When he was in the ninth grade, TJ fell in love with computing. As an adult, he got a master's degree in computer sciences from Virginia Polytechnic Institute in Blackberg, Virginia. Upon finishing his master's in 1979, he joined AT&T Bell Labs and later moved to Hewlett-Packard before switching to Silicon Graphics. He joined @Home in August 1996. In the first six days of the job there, he hired 70 people, taking the total number of @Home employees to 170. TJ had a real cult of personality. He crammed @Home with folks like him—good-looking former jocks. "All flash all the time" is how former insiders described TJ's crew.

In September 1996, the company rolled out its service in Fremont,

California, a Silicon Valley suburb. Within days, the service was a hit. The @Home trucks were being stopped on the street by eager customers who were desperate for the service, @Home chief financial officer Kenneth Goldman later told *Business Week*.[10] The service's scarcity only added to its mystique. "The technology works. The demand is there. The service is great and customers love it. We're at a point where the name of the game is execute, execute, execute," TCI's Ravenel told *Fortune* magazine in November 1996.[11]

There was so much hype around @Home at the time that even Bill Gates of Microsoft wanted a piece of the action; he wanted to merge the struggling Microsoft Network with @Home. "TJ, to his credit, stood right up to these guys at Microsoft. They wanted to put an @Home icon on the desktop instead of the @Home screen most users saw," recalled Sean Doherty, who went to the meeting with Microsoft. TJ's response could be summed up in three words: Take a hike! This was the brilliant side of TJ, a consummate engineer who, like a champion chess player, could see five moves ahead of the competition.

When the service first rolled out in Fremont, the company had an initiative to reach a benchmark of 1,000 customers. "It took us three and a half months to get to 1,000," said Adam Grosser, who was in charge of signing up subscribers at the time. (Grosser is now a partner at the venture capital firm Foundation Capital.) "They had a bell they'd ring each time we'd sign one up. That happened about three times a day."[12] By the end of 1996, @Home had 1,000 subscribers. Not quite the million the company had promised, but on its way there.

By then, the company had moved into its new digs—a two-story, 50,000-square-foot office in Redwood City. Insiders recall that the building, a wide open industrial space, was in terrible shape. The floor was covered with two-square-foot carpet squares, all mismatched, that gave the space a chaotic feel. Insiders joked that the carpet guy was color-blind. They all worried more about asbestos poisoning than about not being able to sign up customers fast enough. "I got there in 1996 and we were working 20 hours a day, because it was such an exciting place," said Lewis Eatherton, a network engineer and one of the key members of @Home's strategic engineering group. "We did not know how to build a cable modem–based network, but we figured it out, and we loved being at work."

In 1997, @Home started prepping itself for an initial public offering. Even though it had about 30,000 customers, for the six months ended June 30, 1997, it posted a loss of around $23 million on sales of $1.8 million. The bankers nevertheless were confident they could sell the loss-making company to investors. But with Internet mania raging, no one cared about these numbers. In July 1997, @Home went public in an offering underwritten by Morgan Stanley, Merrill Lynch, Alex Brown, and Hambrecht & Quist. When the company went public, "we believed that the company would be around forever," recalled Dean Gilbert, former senior vice president and general manager of the company's @Home Group. Other senior executives I chatted with felt the same way.

Investors loved the high-speed service and lapped up the shares of the company just like they did the service. On Wall Street in 1997, broadband was the magic word—it was like saying "open sesame" to wonderful and limitless treasure. @Home raised $94.5 million in the initial public offering. After being priced at $10.50 a share (before splits), the stock jumped 130 percent, and at one point on the opening day was trading at over $25 a share. At the end of the first trading day, the company was valued at $2 billion. Doerr and a bunch of other @Home executives watched @Home make its debut from Morgan Stanley's offices, and after the opening, Doerr remarked, "America's capital markets are a national treasure."[13]

Up to this point, it was all going according to plan—John Doerr's plan, that is. This was the company's honeymoon phase. The investors, the cable companies, and the consumers all loved @Home. The stock closed 1997 at around $12 a share (split adjusted). TJ was frustrated at that point, because the company was now open to scrutiny from Wall Street, which wanted to see quarterly growth. Unless @Home could grow fast, its stock wouldn't rise. However, in TJ's mind, the cable partners were dragging their feet on upgrading their networks to handle the two-way communication that @Home's service required. It was a slow and tedious process, and @Home could do nothing to accelerate it. By the end of 1997, the company had only 50,000 customers—still only 5 percent of the 1 million customers the company proclaimed it would have by the end of 1996. TJ spent most of his time cajoling the cable partners to speed up the upgrades, while the rank and file focused on signing on new customers. By the end of March 1998, the company had boosted its subscribers to 90,000, which helped push the stock to about $17 a share (split adjusted). Revenues were up to $5.8 million for the quarter. But the growth of @Home subscribers came at a cost; the company lost a whopping $95 million in the quarter ending March 31, 1998. More customers meant that @Home needed to bolster its network infrastructure and spend more money. In addition, the company was spending a ton of money on marketing and salaries as well. @Home had launched its service in parts of as many as 26 cities in the United States and Canada.

The growth in subscriber base was taxing the network, which was suddenly overwhelmed. Customers in Fremont, California, and Hartford, Connecticut, started complaining loudly and lobbied local politicians to do something about the declining service from @Home. The complaints scared the living daylights out of the three cable giants.[14] Cable companies are largely dependent on the largesse of the municipalities and local governments, which have a lot of influence in setting prices for services and licenses to offer cable service in a city.

They were worried that their cable licenses could get yanked, and griped about the technology problems at board meetings. At the time, the board was controlled by TCI,[15] and its members included Leo Hindery Jr., then president of TCI, along with John Malone and Bruce Ravenel. Apparently, shouting matches would break out during board meetings. *BusinessWeek* reported that in one of these sessions, Hindery, a cranky, tough-talking cable veteran, traded expletives with Jermoluk.[16] A former executive told me that TJ and Hindery were constantly at loggerheads.

"We're their customers and we never felt they cared about their customers' needs," Dallas Clement, Cox's senior vice president for strategy and development, lamented.[17] Cable guys thought that they were @Home customers—which they were—while the @Home crew thought the consumers were their real customers. These were ominous signs of a deteriorating relationship with the cable companies.

Hollywood Dreams

By the end of 1998, the stock market was in the grip of dot-com mania, and investors were valuing Internet stocks like Yahoo! and Amazon.com much higher than they were brick-and-mortar companies such as Barnes and Noble and Time Warner. America Online, another Internet high-flyer, was being valued at $70 billion in market valuation. It seemed that content was king, and Jermoluk didn't like that a wee bit. The @Home stock, while supremely overvalued at about $31 a share, lacked the pizzazz of other Internet stocks.

According to many insiders I spoke with, TJ was getting increasingly frustrated with cable companies, which weren't upgrading their cable infrastructure fast enough. "TJ was sitting at KP's office all the time, waiting to talk to John," said Doherty. The cable companies' inaction was driving him up the wall. Without such an infrastructure improvement, high-speed Internet access couldn't be rolled out, thus impeding the growth of the whole network. Ironically, the biggest culprit was TCI, the single largest shareholder of @Home. TCI, which had announced that it was going to merge with AT&T in June 1998, had a reason to slow the upgrades. Its networks were in terrible condition, and its service was lousy. John Malone wanted to flip this network of dilapidated assets to AT&T, hoping that it would deal with the upgrades. He was getting rich, he wanted out, and @Home was tangential to his end goals.

Whether it was the stock market's love affair with everything dot-com or his increasing frustration with the cable companies, TJ decided that @Home needed a media and content strategy. He wanted @Home to be a media company—an entertainment network for the new millennium. The idea was to team up with content providers like MTV and charge premium prices for that content.

TJ believed that the very premise of @Home—Internet access over the cable pipes—was going to get commoditized. For a company that was bleeding millions of dollars in cash, this possibility was like a death knell. The company had raised $485 million in debt at the end of 1998 and was still struggling to resolve the technology issues that were plaguing the network. Instead of focusing on improving its core business, TJ decided that in 1999, @Home was going to become a media company. And it would also help the company get out from under the yoke of the cable guys.

"TJ wanted us to be another HBO, and to me it was clear that the CEO did not like his job," recalled one insider, who vehemently opposed the idea of turning @Home into a content/media company. He wanted to give more stakes to the cable guys and ensure that the company had preferred access status for the longest time. "They (cable companies) were our customers, and we needed to keep them happy," the insider said. What is hard to understand is why a company with a near-monopoly on the high-speed Internet access business would change its business model. Externally, the company was feeling the pressure. Rivals like AOL were going to court and trying to get @Home to open up its network so that consumers could order AOL over the cable pipes. One thing that was going right for @Home was that its stock price was rising faster than mercury on a hot summer day in Manhattan. By the end of 1998, @Home's stock was trading at $37.13 a share and rising.

Party Like It's 1999

On January 19, 1999, @Home announced that it was buying Excite, an also-ran Web portal, in a $6.7 billion stock swap. "We are merging with Excite not only for what they have achieved, but what we become together—the new media network for the 21st century," Jermoluk boasted in a press release.

Wall Street was impressed, but inside the company everyone was shocked. There were rumors that the deal was done under pressure from Kleiner Perkins, which had also backed Excite. A popular conspiracy theory was that the deal was structured so that Kleiner Perkins could distribute its Excite holdings to its limited partners (normally university endowments and pension funds). However, it is unlikely that the deal could have been done without the blessings of the board of directors, which was weighed down with cable company executives.

Excite was a Web portal that had fallen behind its main rival, Yahoo!, in the Web traffic and eyeball sweepstakes. The company had been started by computer math majors Joe Kraus, Graham Spencer, Mark Van Haren, Ben Lutch, Martin Reinfried, and Ryan McIntyre from Stanford University in February 1993. The group had developed a search algorithm that could do complex searches, and had gone to Kleiner Perkins Caufield & Byers to raise money. Partner Vinod Khosla gave them $5,000 to buy a big disk drive, try out a large-scale search, and see if it could work on the Web. It did, and that helped the guys get funding from Kleiner Perkins and another top venture capital firm, Institutional

Venture Partners, in 1994. The company changed its name to Excite (from Architext) and jumped into the Web waters, only to be trumped by the likes of Yahoo!, which had adopted the "portal" model early on. Free e-mail services and news headlines made Yahoo! a bigger hit with Web users.

So when @Home came calling, Excite CEO George Bell decided to sell out. The deal was not that well received by some board members. Leo Hindery, the cantankerous board member who was now representing AT&T (TCI closed its merger with AT&T in 1999), was publicly questioning the reasoning behind the merger, and was upset about @Home tying so closely with a single source of information.

The Excite@Home merger, which was technically an acquisition of Excite by @Home, and which closed in May 1999, was a culture clash. Many believe that was the beginning of the end for both the companies. "We were network geeks and there we were merging with Excite, which was all pretty girls, clicks, and eyeballs," recalled Lewis Eatherton, a former employee. Soon after, instead of talking about networks, all discussions were about page-views and click-throughs, he lamented.

A former Excite employee pointed out that Excite typified a lot of the stereotypes of pre-crash Internet companies. There was a lot of hubris and Excite was very image-conscious. Post-merger, however, the only way to describe the company was complete chaos. "The two cultures definitely did *not* mix well. At the beginning, the Excite side of the house really was dominant, as the vision to challenge AOL required getting quality media from Excite down the pipes of the @Home customers," said this former Excite insider, and added, "I would say that the company (especially the Excite side) was very spin-conscious, and was always trying to project the image of a media titan. There certainly were a lot of lavish events, but I always thought it was something to be expected given the perceived scale of the company."

Jermoluk was completely disengaged after the merger. Bell, Excite's chief executive, who was also the new company Excite@Home's CEO designate, was busy commuting between coasts and was a poor leader anyway, according to insiders. The rest of the senior staff were busy angling for top positions in the newly merged company.

TJ was much better as a visionary and a relationship builder than as a CEO focused on execution. He was perfect for the first several years of

@Home, when he had to pitch the vision to the cable partners and get them to work together, and when he had to pitch Wall Street to get cash in the door. He also ably inspired the troops to work long hours. After the merger, there were too many cooks in the kitchen, and the cable partners apparently stopped cooperating with Excite@Home and each other. Once that happened, "TJ's job was more or less rendered impossible; he couldn't sell the vision of the combined company to the cable partners and lost the confidence of the board," recalled a former executive.

Many said the problem was that TJ was all about TJ—he would never give employees due credit. "We all thought an alliance with Yahoo! would make more sense, but TJ wanted something else," said a former executive. "He is a smart guy, but his personality and ego and motivation colors his views of people and the world." The *New York Times* caught wind of the brewing crisis and brought the dirty laundry out in public. "A Hitch to Marital Web Bliss: Excite@Home is often at odds with its cable parents," the *Times* reported on June 9, 1999. So far, the story had been relegated to online news services such as News.com, but the *Times* took it national. "We think that over time, the revenue from transporting data will continue to fall. That's why in our long-term business model, half our revenue comes from the media side," TJ told the *New York Times*.[18] To say that the executives at Cox, Comcast, and AT&T (TCI) were pissed off by that would be an understatement, some former executives revealed.

The stock of @Home, which had hit an all-time high of around $95 a share on April 12, 1999, started skidding and never really stopped. By June 1999, the stock had sunk to the low-$50s and was headed further south. @Home's media moves were beginning to antagonize its cable partners, who were clamoring to get the networks fixed and working. No one really knew the extent of the internal strife between @Home and its cable partners.

The *New York Times* story damaged the already fragile relationship between the cable guys and TJ's crew. And more troubles were to follow. The Excite@Home merger invited unwanted attention from regulators and from the government, as other Internet providers like America Online started lobbying for open access, saying they too should be able to use the cable lines to service their own customers. The whole issue of open access was a political hairball. At a city hall meeting in San Francisco,

Excite@Home folks were lobbying the city to help open the cable lines more quickly. One of the weirdest things about the meeting was that the AOL group brought in a bunch of elderly Chinese people wearing hats that said something about stopping discrimination, but it was really just a ploy to win support for their side. At the meeting, groups of Excite@Home employees complained that they lived in San Francisco and couldn't get service from their own company yet.[19] The whole company and its cable partners had to spend time, money, and energy trying to fight back this "open access" attack.

"It was all klieg lights and flashbulbs. Here it was, a $7 billion deal, largest deal on the web, broadband and narrowband coming together. And very quickly a bunch of things went wrong for us," recalled George Bell.[20] The sparring continued for a while. "I have said all along, subscriptions will go to nothing, ISPs will be free, and if that's your only source of income, you risk being leveraged out by people who can combine access with other sources of income," Jermoluk said.[21]

"If there's one thing that you need it's that your team needs to be singing off the same page of music. And that's never been the case with this company," Jermoluk later told *News.com*, an online technology news publication.[22] In the summer of 1999, Excite@Home doubled down on the content and invested about $60 million in content start-ups, including the now defunct Quokka Sports, a sports information web site. These investments were made at a time when the company was financially constrained. For the quarter ending July 31, 1999, the company had sunk into a sea of red, having lost $217.9 million on sales of $70.5 million. Total number of subscribers: 620,000.[23] In a press briefing in August 1999, Jermoluk claimed, "We are definitely supply-constrained right now. We know the demand is out there. The problem is the installers—there are only so many backhoes digging up the streets."[24] By then, TJ had formally handed over the reigns to George Bell, and was never around, insiders said.

Bell comes from a prominent Philadelphia family—his grandfather was Pennsylvania's governor and chief justice of the state's Supreme Court. His father was the co-chairman of the investment bank Janney Montgomery Scott and is a close friend of former President George Bush (senior). Bell attended Harvard University and then went on to become a writer, documentary filmmaker, and television producer. He worked at places such as

ABC and won four Emmy awards. He was working at *Times Mirror* magazines before he was recruited to run Excite, when it was trying to become a "new media" company. He had little or no idea about how broadband—a very technology-intensive business—really worked.

One of Bell's moves was to buy BlueMountain.com, an online greeting card company with tons of users but no revenues. In October 1999, the company paid a total of $780 million—$430 million in stock and the rest in cash—to buy the online company. BlueMountain.com was the offshoot of an operation started in 1971 by Stephen Schutz, a physicist, and Susan Polis Schutz, a poet, to sell poetry and art posters. The Boulder, Colorado, company launched an electronic greeting card business on a whim, when the Internet became popular in 1996. Consumers loved the ability to send greeting cards over the Web, for free, and BlueMountain.com became the 14th most visited site on the Web. Bell's reasoning for the purchase was that Blue Mountain Arts' traffic would add more oomph to Excite, which was falling behind Yahoo! by the month. This was news in an industry that awaited monthly Internet traffic updates the way television executives await ratings reports during sweeps months. In a news release, Excite@Home said that the acquisition would boost its reach to about 34 percent of Web users.

The move caused an uproar within the company, and the board of directors was up in arms.[25] Excite@Home was precariously low on funds, and here was Bell blowing the cash. The situation was so out of control that Hindery quit the board to go work as the chief executive of Global Crossing, leaving TJ and his crew to their own devices. Replacing him was AT&T's Dan Somers, who was not as vocal as Hindery.

In November 1999, barely nine months after the merger was announced, Excite@Home announced that it would issue a separate tracking stock for its Excite division. This was an effort to placate the concerns of the cable partners. "It's a structure to do our thing on the content side of the business without impacting our partners," said Jermoluk.[26] So why did they buy Excite in the first place? The company had to raise another $500 million in debt, mostly to complete its network, and its total losses at the end of 1999 stood at $1.6 billion.[27]

Despite all the boasting, and after running through billions of dollars, as of January 2000, the company had just surpassed 1 million customers—three years behind schedule.

It was time for TJ to sail into the sunset. TJ didn't want to wait around to see the whole thing come tumbling down. He had almost $400 million in stock. In January 2000, Jermoluk said adios to Excite@Home. And he did it in style. To announce he would step down from the company's CEO post but would stay on as chairman, Jermoluk called into an all-company meeting from somewhere off the Chilean coast aboard the *Hyperion*, a 155-foot yacht owned by Jim Clark, the Netscape founder and also TJ's former boss at Silicon Graphics. "TJ checked out when he had to deal with adult issues—it was not part of his DNA," said a former executive. But before he did that, he signed an extension deal with Cox and Comcast, which gave them an exit from their contract with Excite@Home. This was like kicking the sick puppy while it was down.

Champagne Dreams

Excite@Home, meanwhile, kept blowing up its money. "There was a party culture that existed in the company," recalled a former executive, lamenting, "There was a perception of double standards on how the company spent its money." But most of the spoils went to the senior management; the rest of the company never got to taste the good life. The only time the rank and file shared in the fun was when the company had an IPO party in San Francisco, or at the Christmas parties held in posh locations.

It was reportedly spending $3 million per month on more than 500,000 square feet of real estate, including $10,000 per month for warehouse space to house an expensive Jumbotron (TV) Screen that was never used. The screen was supposed to go up on highway 101, right next to Excite@Home's swanky new offices, and would constantly show the Excite@Home web site.

Then the inevitable happened—the dot-com bubble popped. On March 14, 2000, the NASDAQ slid almost 200 points. This was a clear sign that things were going to get tougher for the dot-coms, which were among the heavy advertisers on the Excite@Home network and were thus a major contributor to its revenues. Worried about its investment, on March 29, 2000, AT&T took full control of Excite@Home, with almost 74 percent of voting stock. The plans for its Excite tracking stock were shelved. AT&T got two of the other board members, Comcast and Cox, to step down from the board. In exchange, the two companies, which each owned about 8 percent of Excite@Home or about 30 million shares, handed their voting rights to AT&T. In exchange AT&T assured Cox and Comcast that should Excite@Home's stock fall below $48 a share, AT&T would buy their stakes in the company. That sent morale plunging just like the stock.

In April 2000, George Bell announced a new broadband strategy, but it was too little, too late. In May 2000, TJ resigned as the chairman of the company and joined Kleiner Perkins as a general partner. In June, Excite@Home was trading at around $19 a share. Despite a fancy 16,000-mile fiber-optic network, Excite@Home was under attack from all sides.

Bell's decision to double the bets on broadband was costing Excite@Home big money. Its cash and short-term investments fell to $200.8 million at the end of 2000, versus the $502 million the company had in April that year when Bell announced that the company was making broadband access a priority.

This was a multimillion-dollar effort, and it was plagued with technical problems from the start. Excite@Home had developed a specialized start-up screen—a web page that only Excite@Home's broadband users could see. The service never worked when it was launched in April 2000. Then in June 2000, more than 1.5 million Excite@Home customers experienced e-mail problems, which caused a massive uproar and was covered extensively in the press. In July 2000, Excite.com's web site crashed, and in the fall of 2000, service outages plagued the Excite@Home network. In November that year customers were taking another hit: e-mails were getting lost. No wonder they were complaining—the much-vaunted Excite@Home network was simply falling apart.

Using a competing technology called Digital Subscriber Line, or DSL, the Baby Bells were offering their broadband service. While not as fast as Excite@Home's network, the DSL lines were proving to be more reliable than the coaxial cables. In addition, unlike an upstart dot-com, those companies were established and usually already had long-term relationships with potential customers. There was nothing that could stem Excite@Home's slide. Shareholders were up in arms, as were the cable partners. By the fall of 2000 it was clear that Excite@Home would have to tap the markets again to stay solvent. In September 2000, Bell announced that he was quitting, but added that he would hang around until the company found a new chief executive. Excite@Home's executive suite started resembling that of Global Crossing's, as other senior executives also headed for the nearest exit!

Ironically, at the very time the company was plunging into an abyss, Jermoluk was mentioned in a *Forbes* magazine article about executives who found post-corporate happiness. "Five years into the Internet revolution, Jermoluk and other new-wave Rockefellers are converting stock options into goose bumps—using their newfound riches not to buy pricey toys, but to deepen the experiences and interests they already have."[28] The *Forbes* story talked about TJ's $200,000 stunt plane, *Juliette*. TJ and his buddy Jim Clark were investing in Redwood City, California–based Kibu.com, a Web destination for teenage girls, which shut down just five months after it was launched. To @Home employees it was just unfair.

A Year of Discontent

In January 2001, Bell hired Hossein Eslambolchi, a networking veteran from AT&T. With more than a decade of experience and 100 patents to his name, Eslambolchi was brought in to fix the company's technology problems. He spent almost $54 million on new equipment and successfully improved the network's performance. But that carefree spending pushed Excite@Home even closer to the brink. For the first quarter of 2001, Excite@Home lost $61.6 million, on sales of $142.8 million. In a conference call with investors in April 2001, Bell announced that the company would fall short of its targets and would need an additional $75 to $80 million by the end of June 30, 2001, in order to keep its doors open.

A few days later, having delivered the bad news, the company hired Patti Hart, a former Sprint executive, to be the new chief executive of Excite@Home. Hart had not even gotten comfortable in her CEO chair when she got the bad news that others had expected for months: In six months Cox and Comcast would end their exclusive deal with Excite@Home. That meant Excite@Home would lose exclusive access to Comcast and Cox customers, and the two cable companies could offer their own rival service (which they later did). At that point, desperate, Hart went looking for money from unconventional sources. AT&T bought certain assets of Excite@Home for $85 million. The company raised $100 million from the New York–based Promethean Capital Group, an investment group that acts like a loan shark for companies in deep financial trouble.

Having spent millions on the network upgrades, and following the customer loss to Comcast and Cox, the company was seriously strapped for cash. The stock market had soured on Internet stocks and wasn't a viable option to raise more cash. In July 2001, Hart had to hit the road again to raise more money. The cable companies who owed Excite@Home $50 million did not pay up on time, further adding to the company's misery. But, alas, even if they paid, it would not be enough. Hart pleaded with AT&T's head honcho, Michael Armstrong, for help, but none was forthcoming.[29] It was only a matter of time before Excite@Home filed for bankruptcy, which it did on September 28, 2001.

After the bankruptcy filing, AT&T offered to buy the assets of the failed carrier for $307 million, but those talks broke down, and AT&T withdrew its offer. AT&T proceeded to transfer all its Excite@Home customers to its own network. On February 28, 2002, the company finally closed its doors forever. Since then, bondholders and other investors have sued AT&T, alleging fraud. The lawsuit claims AT&T executives stole Excite@Home's intellectual property. The lawsuit is based on interviews with ex-employees, including its last CEO, Patti Hart.

So here is the final tally of Excite@Home losers and winners:

Comcast and Cox made over $3 billion.

Kleiner Perkins Caufield & Byers profited to the tune of $575 million.

Tom Jermoluk walked out with $400 million in stock, although it is hard to know how much of that stock he sold.

AT&T lost around $3.3 billion.

Investors lost around $1.25 billion (and this does not include those who bought shares in the company at market prices).

Milo Medin, the networking wizard, unlike Jermoluk, never got rich from the company he had helped start.

In the end, Excite@Home failed not because it was a bad idea or because it had no customers. In fact, at the time it filed for bankruptcy, the company had 4.1 million subscribers and was the most dominant broadband player in America. What happened is that the founders of @Home had a great idea but owned no assets to make it happen, so they had to use other people's networks to run their business. They were fully dependent on those other companies, and their service cost a lot to get going because of installations, equipment, and customer service. When their partners changed their minds and decided to go into the business of providing high-speed Internet access themselves, Excite@Home was screwed.

"I think this company was everything that was wrong with Silicon Valley," said Dean Gilbert, the former @Home senior vice president who is still recovering from his Excite@Home experience. The company had everything going for it. It failed because of ego, stupidity, and bad management. That, in the end, is the difference between good ideas and great companies.

8

TEDDY GETS TAKEN TO THE CLEANERS

Americans have always had a genius for communications. Today our world is being remade yet again by an information revolution.

But this revolution has been held back by outdated laws, designed for a time when there was one phone company, three TV networks, no such thing as a personal computer. Today, with the stroke of a pen, our laws will catch up with our future. We will help to create an open marketplace where competition and innovation can move as quick as light. An industry that is already one-sixth of our entire economy will thrive. It will create opportunity, many more high-wage jobs and better lives for all Americans.

The Interstate Highway Act literally brought Americans closer together. That same spirit of connection and communication is the driving force behind the Telecommunications Act of 1996. The Vice President [Al Gore] in many ways is the father of this legislation because he's worked on it for more than 20 years, since he first began to promote what he called, in the phrase he coined, "The Information Superhighway."

President Bill Clinton, remarks at the signing
ceremony of the Telecommunications Act,
February 8, 1996

■ ■ ■

Theodore "Teddy" Forstmann must rue the day he heard about telecom and all things broadband. In less than five years, his flirtation with broadband has left his carefully built reputation in tatters. Forstmann and his partner, former Merrill Lynch investment banker William Brian Little,[1] were co-founders of Forstmann Little & Company, a private equity firm with $7 billion in assets.[2]

Forstmann's love affair with telecom began late in the summer of 1999, when he invested $1 billion for 12 percent of Cedar Rapids, Iowa–based McLeodUSA. Forstmann Little eventually pumped as much as $3 billion into telecoms at a time when it seemed that nothing could go wrong with the communications sector. But when the telecom bubble burst, all hell broke loose. In the end, Forstmann Little went from a top-notch buyout powerhouse to an investment firm being used as a piñata for corporate abuse by the Connecticut state attorney general.

Teddy Forstmann, a prominent New York financier, was born in 1940 as the second child of six to Julius and Dorothy Forstmann. He grew up in extremely privileged surroundings in Greenwich, Connecticut.[3] His father came from a line of German Lutherans who resided in Upper Montclair, New Jersey. Teddy attended Yale University and then Columbia Law School. While he was at law school, Teddy's father passed away and, due to some financial setbacks that followed, his privileged life came to an end.

After meandering through a variety of jobs, Teddy found his bearings when he was in his mid-30s. In 1978 he started Forstmann Little & Company with William Little, and the firm quickly became a dominant force in the world of leveraged buyouts and private equity. Over the next 20 years, with ruthless competitiveness and chutzpah, the two built Forstmann Little & Company into a financing superpower. Teddy (with William) orchestrated buyouts of airplane maker Gulfstream Aerospace and soft-drink maker Dr. Pepper and turned those companies around even when the odds were against him.

Between 1980 and 1992, his firm invested in 18 companies. The limited partners of Forstmann Little, which were typically large pension funds and university endowments, received about $4.9 billion from an investment of around $1 billion over this period, the *New York Times* reported.[4] In the 1980s, Forstmann's public scorn for junk bonds resulted in a long rivalry with the king of leveraged buyouts, Kohlberg Kravis & Roberts, whose antics were well documented in tomes like *Barbarians at the Gate, Den of Thieves,* and *The Predators' Ball.*

Despite its decade-long success, Forstmann Little started to slip into the background in the early 1990s, although there would be occasional mentions in the society pages. Forstmann was linked with the late Princess Diana[5] and with model Elizabeth Hurley. He would also get some press for his many philanthropic activities. His close friends include the current Secretary of Defense Donald Rumsfeld, who, like Teddy, is a member of the board of directors of Empower America, a conservative Republican group open only to the elite, according to the *New York Times.* Teddy was the founding chairman of the group.

Private equity funds like those run by Forstmann and Tom Hicks of Hicks, Muse, Tate & Furst typically buy firms that generate a good deal of cash flow. Normally, they find buyout opportunities in ailing grocery store chains, auto chains, and other businesses that generate a lot of cash but not necessarily much profit. Their buyout targets are often publicly traded, undervalued, with little or no debt, and in need of fixing. Usually, the leveraged buyout guys would load up the company with debt, cut costs, try to fix the operation, and then sell it. Many used junk bonds to finance these companies, but Forstmann Little, notably, stayed away from junk bonds and instead used real, hard cash and other more prudent financial instruments. But from 1994 on, U.S. markets started rallying, and stock prices were rising faster than hemlines at Paris couture shows. That made finding bargains even in industrial America harder because the stocks became too expensive. And this made their primary job—making money for their investors—even harder.

Then came the Telecommunications Act of 1996, and it changed everything.

Act Now or Die

It can be said now that the birthday of the telecom bubble was February 8, 1996. On that day, with the stroke of a pen, President Bill Clinton turned the normally staid world of telecommunications into the Wild West. The new Telecommunications Act—the handiwork of, among others, a Democratic vice president, Al Gore, and a Republican house speaker, Newt Gingrich—promised to unleash competition in the erstwhile closed phone industry, help create new phone companies, and create a world in which high-speed Internet access would be a norm, not an anomaly.

"The ultimate judge of our competition policy is the capital markets. If they don't invest in competition, we won't get competition," Reed Hundt, then the chairman of the Federal Communications Commission (FCC), told the *New York Times* at the time. Political leaders urged Americans to participate in a historic opportunity. Senator Larry Pressler, Republican from South Dakota and the chief Senate sponsor of the legislation, reportedly told Clinton that the signing of the bill "was like a gun going off in the Oklahoma land rush. There will be an explosion of new investment in our country."[6] In essence, Washington was telling the rest of the country to go forth and speculate on the brave new world of broadband. Beltway gurus promised an era of endless prosperity. House Speaker Newt Gingrich called it a jobs bill, claiming that the overhaul of the telecommunications industry could produce 3.4 million new jobs.

The act urged Baby Bells and other large phone companies like AT&T to open up their networks to new competitors. Such was the level of optimism that everyone wanted a piece of the action—within three years, there were over 500 carriers, up from 243 at the time the act was passed. During the five years since the act was passed, every year brought forth new business models and fancy acronyms. These new age carriers, the CLECs, DLECs, BLECs, and DSL, should have been collectively referred to as NFP: no frigging profits.

The dominant mindset was this: Take two former Bell employees, give them a laptop, fire up Microsoft PowerPoint, whip up a presentation, hit "control" and "S" to save, and raise a few million dollars to start a company that could offer high-speed Internet access and phone service over the same connection.

Death in the Slow Lane

It all started when a couple of Intel executives figured out that offering high-speed Internet access over existing phone lines could prove to be a gold mine. Thus began the great craze for digital subscriber line (DSL) technology that enabled sending data over plain old phone lines at super high speeds. The idea was the brainchild of a telecom attorney, Dhruv Khanna. Khanna worked for Intel in 1993 and helped the chip giant understand how deregulation of telecom would influence Intel's primary business of selling chips that power a majority of the world's personal computers.

At the time, Charles Hass, another Intel employee and a friend of Khanna's, was helping Intel identify promising data access and transport equipment makers. Hass and Khanna quickly realized that DSL technology could provide the means to their own fortunes. Khanna, as an attorney, had already managed to wrangle his way into Washington's power circles and was actively involved in the final writing of the 1996 Telecom Act.

Khanna and Hass realized that the Telecommunications Act of 1996 would force Baby Bells to share their local access, or "last mile" infrastructure, with all newcomers. In October 1996, the duo quit Intel and started Covad Communications, the first of many DSL companies that would be founded in the first 24 months of the signing of the Telecom Act of 1996. Intel invested $1 million in seed capital and helped the company attract another $7.5 million from other venture capital investors. Charles McGinn, another Intel veteran, joined the company as the chief executive. Soon, others jumped on the DSL bandwagon. NorthPoint Communications, Rhythms NetConnections, Telocity, FlashCom, and several other DSL companies set up shop between 1996 and 1998.

As one venture capitalist told me, the whole DSL craze started because the venture capitalists were sick and tired of getting on-line with their infuriatingly slow dial-up modems. The idea behind offering DSL service was quite smart, actually, at least from a business perspective.

The small- and medium-sized businesses were underserved by the Baby Bells. These businesses could not afford the $1,500 a month charge per T-1 line offered by the large phone companies. At $79 to $150 per month, DSL lines were much more affordable. But it was not going to work according to the start-ups' best-laid plans. Since the DSL threatened to decimate their highly lucrative T-1 business, large phone companies like Bell Atlantic and NYNEX (now Verizon) started dragging their feet and blocking these upstarts from getting access to their central offices and phone lines. Bells were simply exhibiting good business sense—protecting their business and the interests of their shareholders.

And if that was not enough, many of these start-ups made stupid mistakes of their own. For much of 1997 and 1998, the scope of the big three DSL companies—Covad, Rhythms, and NorthPoint—was limited. They were going to be wholesalers who sold to more marketing-oriented DSL companies like FlashCom and Telocity, which in turn would sell it to end customers. It was a simple food chain.

The problems for the big three started after they went public. Wall Street wanted to see revenue growth. Since none of these companies had any profits, the analysts started to use number of customers as their growth metric. In order to meet the Wall Street expectations, the DSL providers built expensive national infrastructure and quickly saturated the market. When that stopped working, they entered the retail market. Not only was this an expensive undertaking, it also alienated them from their original customers, medium-sized Internet service providers who were their resellers, and cost them millions of dollars in marketing and sales costs. Claudia Bacco, president of the consultant group Telechoice, believes that the retail strategy is what really pushed the DSL companies over the edge.

Using funds raised from their public offerings, these companies brashly built national footprints but never managed to get enough customers, in the same "build it and they will come" spirit of the fiber barons. The markets paid no heed and valued these companies at hyper-inflated levels, despite their nominal customer base and underlying difficulties obtaining access to the "last mile."

The CLEC Madness

Another type of company that sprouted after the Telecom Act of 1996 was the competitive local exchange carrier, or the CLEC. And like the DSL companies, CLECs like Allegiance Telecom, Time Warner Telecom, Focal Communications, and McLeodUSA also tapped big-ticket investment funds to develop their capital-intensive infrastructure.

A CLEC provides local, long distance, and Internet services largely to business customers. In order to do that, they need to share the Baby Bells' central offices—a warehouse-style space where all the local lines from a neighborhood terminate. The Bells owned this warehouse space, but the Telecommunications Act of 1996 forced the Bells to share this space with the upstart CLECs. The only consolation was that they could charge CLECs for the privilege. This space-sharing arrangement was called co-location. A CLEC that wanted to offer service in a certain city had to share the space with the local phone company, because the phone calls had to be transported over the Baby Bell's phone lines to the CLEC's equipment, and then to the back-haul network that connected different cities and countries, where it would be carried to its final destination.

According to some estimates, a typical CLEC would pay one-time fees of around $60,000 for 100 square feet of space in a Baby Bell central office. Then it would pay around $5,000 a month on top of that in other fees and service charges. In order to serve about 1,000 customers, the CLEC would typically need 350 co-location centers. This meant that even before it signed up a single customer, the CLEC was spending around $21 million to get its co-location facilities in place.[7] (And this did not include the cost of equipment.) These up-front capital costs made the business model more challenging (even economically unfeasible), and it forced companies to take on a load of high-interest debt. Most CLECs didn't have a choice, because no one would want to do business with small operations like theirs unless they provided service on a par with the likes of Verizon and SBC Communications. And that required building their own nationwide network, even if it cost them billions of dollars. Lucky for them, there were enough punch-drunk investors walking around with open checkbooks!

By the end of 1998, venture capital firms active in investing money in the CLECs included Battery Partners, Crescendo Ventures, Crosspoint Ventures, Madison Dearborn Partners, Patricof & Company, Centennial, Columbia Capital, Spectrum Equity Partners, and M/C Venture Partners. Typical investment rounds were around $100 million, often consisting of blended equity and debt investments.

"We invested in a lot of companies—ISPs, CLECs, hosting companies, BLECs—because it was an attractive space," said David Sprague, general partner with Palo Alto–based Crescendo Ventures. Sprague said that one of the reasons they invested in these companies was because they believed Internet traffic was doubling every 100 days. (Of course, that wasn't exactly the case—see Chapter 1.) In addition, many believed that CLECs, using superior technology, could beat the Baby Bells on price and therefore steal market share. Many underestimated the build-out costs, and most were hoodwinked by revenue-growth falsehoods being perpetuated by WorldCom. "They were too capital intensive, and I think we took a lot of financial risk—we were early stage guys and we were putting in $100 million when the companies needed $700 million," said Sprague.

These investments were not enough, and pretty soon the companies needed more money. Traditional venture capital firms didn't have the kind of money these start-ups needed. So the start-ups went looking for money elsewhere—in this case, from the private equity world (and public equity markets) or from the debt markets via mega junk-bond offerings. These high-interest bonds, which were, in essence, nothing but blood sucking IOUs, were sold to mutual funds, insurance funds, and pension funds.

Meanwhile, buyout firms like Forstmann Little had been hobbled by rising stock markets that had made most of their traditional targets—largely industrial era companies—more expensive. These firms watched their Silicon Valley counterparts, the venture capitalists, making billions of dollars from their technology and telecom investments. These venture firms were threatening to become the most-favored investment tool for limited partners.

The Wall Street bankers, too, were well aware of the crisis facing the buyout business. The buyout kings had few deals, and they were charging almost 2 percent annually in management fees on funds that were about $1 to $3 billion in size. They were under increasing pressure from their limited partners to show decent rates of return on investments. Wall Street knew that eventually the likes of Forstmann would fall for the telecom chimera—it was only a matter of time. Portfolio managers of the bond funds faced the same predicament, as investors fled their funds to invest in the high-growth technology equity funds.

Capitalizing on the desperation of both the parties—sellers (broadband start-ups) and buyers (in this case, the buyout crew and bond investors)—was going to result in the sort of investment banking fees that still fuels new home building in the Hamptons. All Wall Street had to do was package the new companies in a way that would make buyout guys whip out their checkbooks. And that was easy—the telecom stocks were going to the moon, and that was sufficient for most investors.

At most Wall Street banks (like Merrill Lynch, Morgan Stanley, and Goldman Sachs), the senior managing directors focused their energies on large blue-chip telecom clients like AT&T, Bell Atlantic (now Verizon), NYNEX (now part of Verizon), Ameritech (later acquired by SBC Communications), Pacific Bell (acquired by SBC), and Bell South. These accounts were handled by the very senior folks, and many second- and junior-level bankers never got to see the gravy.

Frustrated, the young turks turned their attention to shakier CLECs and aggressively courted their business. Many of these junior bankers had little or no knowledge of the telecom industry. If quizzed, they probably couldn't tell the difference between SDSL and ADSL—two different flavors of DSL technologies.

"We completely manipulated the stats so that the company looked good, and played up our own strengths," a former telecom banker said, wishing to remain anonymous for obvious reasons. "Our pitch books had nothing much to say, but we encountered companies that were hungry for cash, and (these) people believed that the profits would take care of themselves. The belief was to build the operation, and sell it to someone." Getting sold to WorldCom was seen as the ultimate exit. Investment bankers who brought in about $100 million in investment banking fees for the firm could easily go home with more than $5 million a year. "We came up with the deals and the sales force sold them," said this former banker. "Bankers didn't have the best interest of the company at their heart. It was all hail to the mighty dollar."

Small wonder then, that investment banking fees (includes debt, convertibles, and equity-related issues) for the telecom sector increased from a mere $1.06 billion in 1996 to $4.14 billion in 2000, according to Thomson Financial data. Topping the charts were the big five—Salomon Smith Barney, Morgan Stanley, Credit Suisse First Boston, Merrill Lynch, and Lehman Brothers. Most of these deals got done because high-yield debt buyers were willing to buy anything telecom. Through 1998 and 1999, the stocks of CLECs and DSL companies were trading at sky-high valuations.

"We think this is a huge growth sector within the telecommunications industry. Essentially the 1996 Telecommunications Act opened up a

$100 billion market to competition," Simon Flannery of J. P. Morgan Securities told *Red Herring*.[8] This was a very compelling argument when it came to selling junk bond deals. All investors in the CLECs were thinking that if one of the telecom companies they backed could capture even half a percent of the total market, they would be sitting pretty.

The rub was that there were too many telecom companies trying to raise money, so it was tough to sell bond offerings without some sweeteners to potential buyers. The high-yield bonds (read, junk bonds) would have to carry really high interest rates, in some cases around 10 percent, and the bonds would have to have warrants attached to them, to allow buyers to profit from any upswing in the stock price. A warrant is a special financial instrument that allows its holder to purchase a company's common stock at a fixed price. Another prevalent practice, according to bankers immersed in those deals, was to require that companies set aside about 30 percent of the money they had raised to meet interest payments on the bonds.

Say a company raised $100 million and then put 30 percent aside to guarantee that its debt owners got interest payments over the next three years. While the company had raised $100 million, it could use only $70 million but had to pay back $100 million plus an additional $30 million in interest over three to eight years, depending on the type of offering. The interest rate on these bonds was typically between 8 and 12 percent. The debt holders who were buying the bonds were basically assured of a 10 percent rate of return every year. And that was enough to make them look good to their investors. With some of their downside protected, the bond buyers threw caution to the winds and started buying a lot of high-yield bonds. Ravi Malik (no relation to this author), a portfolio manager at the bond investment fund Froley Revy, told me in an interview that bond offerings were so numerous that from 1998 through 2000, there would be someone pitching such bonds to them every week.

Private equity investors proved to be equally carefree. Unlike venture capitalists, who typically invest in very risky early stage companies or ideas, the private equity investors invest in mature operations or do leveraged buyouts. But in 1999, the buyout funds that had been watching the frenzy from the sidelines also finally caved in and started opening their multbillion-dollar war chests to become the primary financiers of the last-mile revolution. "The private equity people lost their focus and moved into the venture capital world," said Harvey Miller of private equity firm Greenhill & Company, in an interview with *Investment Dealers' Digest.*[9]

Investments in the telecom services business and Internet service provider businesses shot up tremendously in 1999, and kept rising through 2000 before sinking like a rock in 2001. The funding for U.S. broadband companies ballooned to $4.25 billion in 1999 and topped out at around $6.7 billion in 2000. Frighteningly, the euphoria in the United States was mirrored through much of the developed world.

After DSL, it was CLECs, and then came the BLECs, or Building-based Local Exchange Carriers. BLECs are service providers that bring broadband and voice connectivity to office buildings, industrial parks, hotels, and, in some cases, apartment buildings. Many of the BLECs, like Allied Riser Communications, Cypress Communications, and Broadband Office, raised more than $250 million each from private investors, through late 1999 and early 2000. However, their funding paled in comparison with the next generation of CLECs: the metro area network providers like Yipes, Telseon, and Sigma Networks, whose chairman, incidentally, was Reed Hundt, the former chairman of the Federal Communications Commission. The metro service providers alone raised more than $1.25 billion in 2000 and 2001. Even at recent count, the metro services market has a mere $400 million in annual sales. (Many start-ups, including Sigma, Telseon, and Yipes, went belly-up later, with Cogent and OnFiber the only survivors so far.)

The cavalier nature of funding for these companies was reflected in this statement by a prominent venture capitalist, Peter Wagner, a partner with Accel Partners, who in March 2000 told *Venture Capital Journal*: "The big difference now in telecommunications is a much more open and freewheeling industry structure involving the Internet. Literally anybody can hook up to a piece of equipment on this open network, which is ruled by open protocols, and start innovating what applications or services equipment to build."[10]

Teddy's on McLeod Nine

Teddy Forstmann's firm Forstmann Little was one of those companies funding all things telecom. Forstmann's first $1 billion telecom investment was in McLeodUSA, a company started by Clark McLeod, a former math teacher who many have compared to the late Mr. Rogers, the cardigan-wearing host of a children's television show on PBS. (If that seems odd, remember that Bernie Ebbers was a gym teacher.) McLeod had started a long distance reseller business in 1980 called TelecomUSA, which he sold to MCI in 1990 for $1.3 billion. A year later, he started another company, McLeodUSA, which sold local and long distance service in the American Midwest. After growing gracefully for many years, McLeodUSA went public in the summer of 1996 at $20 a share. The stock rose in tandem with the mania around broadband stocks and anything telecom.

McLeodUSA grew ambitious and planned to build a 6,000-mile fiber network by 2002. It wanted to grow aggressively into other businesses like publishing telephone directories and telemarketing. It was even thinking of buying competitors. To do all that, McLeodUSA needed a larger bankroll, and Forstmann Little smelled an opportunity, even though McLeodUSA already had about $1.8 billion in debt. Forstmann Little ended up buying 12 percent of McLeodUSA. At the time, it seemed there was little to worry about, as the stock was trading in the mid-$30s and the communications mania continued.

A few months later, Forstmann Little decided to take a bigger bite out of the telecom pie. In December 1999, the buyout giant invested $850 million in Nextlink Communications, another local-exchange-carrier-meets-fiber-optic-network company (a souped-up CLEC). Nextlink was the brainchild of Craig McCaw, the wireless wunderkind. In January 2000, Nextlink decided to merge with Concentric Network, an Internet service provider that primarily served small- and medium-sized businesses. Forstmann Little invested another $400 million in Nextlink in May 2000, and the company changed its name to XO Communications. By then, Forstmann's fund was on the hook for $2.25 billion in telecom investments that included XO and McLeodUSA.

In 1999, telecom was the best-performing sector of the stock markets—up 121 percent versus NASDAQ, which, propped up by the new telecom stocks, was up 86 percent, from 1998. By comparison, the Dow Jones Industrial Average was up only 25 percent in 1999. Around this time, other buyout funds also made forays into telecom funds that would eventually prove disastrous. Hicks, Muse, Tate & Furst, which was started by Tom Hicks, owner of the Texas Rangers baseball team and a close friend of President George W. Bush, invested more than $1 billion in ventures like ICG Communications and Winstar Communications. Others, like long-term telecom investors Providence Equity Partners, Welsh Carson, The Blackstone Group, The Carlyle Group, and even hedge funds' private equity arms, got into the action. Many of them were arriving at the party too late. Soon, in March 2000, disaster would strike and the dot-com bubble burst would eventually take CLECs down with it. And the worst was yet to come.

Numbers Don't Add Up

One man, Ravi Suria, had concluded after months of study that it was only a matter of time before these new companies started dropping like flies. Even though some investment banks had cautioned investors, the majority of Wall Street was on a massive telecom binge. In early 2000, Suria was a credit analyst for Lehman Brothers, a white-shoe investment bank. On June 23, 2000, he ruffled more than a few feathers when he put out a report that questioned the viability of Amazon.com, the online retail giant. Unlike others on Wall Street, Suria was sure that Amazon's borrow-and-grow strategy would eventually become its undoing. Suria wrote in a report that, to succeed, Amazon "must be able to generate the cash operating profile of a successful retailer. It is essentially this yardstick that we use to analyze the company and, as the rest of this report shows, we find it woefully lacking." Amazon's bonds skidded 10 percent that day and shares tumbled 19.4 percent, to $33.88. Amazon called the report error-ridden and started maligning Suria.

Wall-Streeters dubbed the then 29-year-old analyst, "Amazon killer." Born in Madras, India, Suria had toiled in relative obscurity up until his run-in with the Amazonians. He had ranked third for three years in the *Institutional Investor* magazine's poll of most influential analysts. (Think of the poll as the People's Choice Awards, where investors like mutual funds and hedge funds get to vote for their favorite analyst.) Suria wanted to get out of the media spotlight and retreated to his small office on the fourteenth floor of Lehman's office at 200 Versey Street. His workspace was cramped between that of a real estate investment trusts (REIT) analyst and a gold analyst, across from the offices of Lehman's technology research group, and five floors above the high-yield and fixed income research group. Lehman, it seemed, didn't much care for Suria.

With all the noise and controversy around Amazon.com, Suria decided to shift gears and started to put the broadband business under his microscope. When most of his peers were summering in the Hamptons or shopping in Milan, Suria read hundreds of SEC filings, analyzing the financials of companies like Winstar, Focal, and NorthPoint. His conclusion: In a matter of months, the purveyors of broadband would either have to start making a ton of money or would simply go broke within months. "It was not that I woke up one day and discovered the telecom market. Since 1997, telecom companies had become one of the biggest borrowers and the amounts just kept increasing, and they became 40 percent of the convertible market I followed," he recalled.

In August 2000, he started working on a telecom report titled "The Other Side of Leverage." (Suria apparently wanted to call it "The Dark Side of Leverage," but changed its title later.) It took him 45 days to write the report and have it approved by the firm for distribution. Suria claims that the firm dragged its feet. After all, in 1999, Lehman had almost $160 million in investment banking fees from the telecommunications sector.

Suria's findings were startling. Since the Telecommunications Act of 1996, new-age telecoms had raised about $213 billion in debt, and another $62 billion in convertible bonds (which are bonds that are convertible into an issuing company's common shares at a pre-set conversion price). This didn't include the debt raised by the likes of AT&T and WorldCom, which totaled a whopping $265 billion. In comparison, in the heyday of junk bonds, between 1983 and 1990, the total debt raised was $160 billion, and the companies that raised those bonds were mostly cash-flow positive.

"Moreover, the debt from the high-yield and convertible markets does not represent the total borrowings of these young companies, as these numbers do not take into account transactions such as vendor financing, syndicated loans, or lines of credit," Suria wrote in this report. Vendor financing (see Chapter 10) was a type of financial skullduggery common during the telecom boom. Essentially, companies like Nortel and Lucent would loan companies hundreds of millions of dollars in exchange for a promise that those companies would turn around and buy equipment from them. In short, it was only a matter of time before the borrowing binge would push the industry into a financial abyss. Suria noted, "The investment thesis for many of these firms was that as they were nimbler and faster than the incumbents, they would quickly raise capital, build out networks with the latest technologies, and then sell the completed networks to the large investment-grade telecom companies."

Essentially, what Suria was saying was that, despite all the talk of hypergrowth, the demand was not there. Most companies had created a false sense of illusion about revenue growth, and this phenomenon later came to be known as multiple counting. For instance, South Ferry Printers buys a T-1 line from Carrier X, and pays Carrier X $1,500 a month. Carrier X in turn leases the line from Baby Bell A and pays Baby Bell A $500 a month. Then Carrier X signs up with Carrier Y to provide the backbone network bandwidth and pays Carrier Y $500 a month. Carrier X was essentially getting $500 a month in revenues, which in most cases was not even enough to cover its costs. However, the three companies showed total monthly revenues that exceeded what the customer actually paid:

$$\text{Carrier X + Baby Bell A + Carrier Y}$$
$$= \$1500 + \$500 + \$500 = \$2500 \text{ a month}$$

—even though the customer only spent $1500!

Replicate this scenario across the country, indeed the world, and the revenue numbers go totally haywire. This sloppy accounting helped companies raise billions from the market, while masking structural problems with the businesses. It was a giant Ponzi scheme. The only way these companies could make their money back for investors was to sell out to a company like WorldCom. Suria had put the whole thing together in black and white.

In October 2000, Suria and Derek Harris, an institutional salesman (and now an executive vice president with Lehman), went to meet many leading institutional funds like Putnam and Scudder Investments, urging them to take a second look at the whole telecom sector. It was a warning, but many of them ignored Suria. One man who paid attention was James Cramer, then a hedge fund manager, who had co-founded TheStreet.com, an online financial news magazine. Cramer realized Suria's thesis had merit. He had lunch with the analyst and later wrote about it. "I could barely contain myself. In 15 years of running money I never saw anyone with discipline and conviction—with the possible exception of Dan Benton when he was at Goldman Sachs—like Suria's," he wrote in his online column for TheStreet.com on March 29, 2001. "This guy is money in the bank, the best young analyst . . . that I have ever met." Whatever Cramer's shortcomings, he has at least one asset—he trades on good ideas when he hears them.

Suria's report once again put him in the eye of the storm. He would make his pitch, only to have it denigrated by the likes of Jack Grubman, the former Salomon Smith Barney analyst who is now under investigation by the Securities and Exchange Commission for his alleged role in the broadband bubble.

The pressures from inside Suria's firm were immense. Since 1996, Lehman was one of the top underwriters of telecom debt and convertible securities, generating enough profits for many of its investment bankers to upgrade their Porsches every year. The report had caught the attention of many big-money managers including George Soros, Stan Druckenmiller, Louis Bacon, and Lee Ainsley—smart money people who rarely talk to street analysts. "And perhaps the most significant thing was that after a breakfast meeting, Druckenmiller offered me a job by lunch to 'come and think and generate moneymaking ideas'—which vastly expanded my role from flogging convertibles," said Suria.

For Suria, it was time to call it quits. In March 2001, he walked out of Lehman's offices and went to work at Stanley Druckenmiller's Duquesne Capital, a hedge fund. He was finally with people who appreciated his skepticism. And as he left Versey Street, he knew that the phone companies, like hyperkinetic teenagers who had maxed out their credit cards, wouldn't know how to pay off the charges. The NASDAQ stood at 1,972 and the Dow Jones Industrial Average was at 9,947. To Suria it was clear—the market and telecoms were going down, and fast.

Bankruptcies Loom Large

Some of Suria's predictions were already coming true. By March 2001, two dozen communications companies had drowned in debt. The first one to run out of air was ICG Communications, which filed for Chapter

11 bankruptcy protection in November 2000. ICG was run by J. Shelby Bryan, a Texan who had moved to New York and since become a big gun on New York's social circuit. He also started raising funds for the Democratic Party and got himself a new girlfriend, Anna Wintour, *Vogue*'s editor in chief. "Bryan was a flashy extrovert with undisguised social ambitions. A country-clubber with a notoriously roving eye, he was becoming a player both in Washington circles and on Manhattan's power scene," wrote *New York* magazine in 1999.[11] The magazine estimated that Bryan was worth around $30 million, including a $3 million home in East Hampton. Bryan had made some money from ICG, but others, like Hicks, Muse, Tate & Furst, Gleacher & Company, and Liberty Media, who had collectively put $750 million into Bryan's company, were not so fortunate.

Despite the bankruptcies, the companies' founders always came out ahead financially, according to a study conducted by the Peace and Freedom Foundation (PFF). Based on data reported to the Securities and Exchange Commission, the PFF study shows that 59 insiders at seven leading competitive telecom firms sold stock worth $1.4 billion, an average of $24 million each, during 2000 and 2001. In comparison, the study reports, the current combined market capitalization of the firms is only about $1 billion. The firms studied included Allegiance, Covad, Focal, McLeodUSA, Global Crossing, Level 3 Communications, and Rhythms (see Figure 8.1).

Meet some of the profit takers:

Catherine Hapka, former CEO of Rhythms NetConnections, sold 800,000 shares for more than $21 million—about 30 percent of her stake—from November 1999 to August 2000. The company's status today: out of business.

Robert Taylor, CEO of Focal Communications, sold about 500,000 shares worth about $7.8 million from June 2000 to April 2001. Today, the company trades at $.45 a share on NASDAQ.

Dhruv Khanna, a founder and executive vice president of Covad Communications, sold about 1 million shares—about 20 percent of his holdings—from January 1999 to January 2001, for about $37 million. The company has recently reemerged from bankruptcy, a pale shadow of its former self.

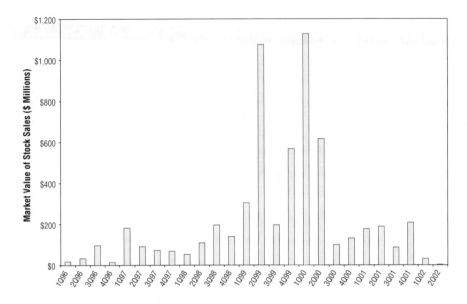

Figure 8.1 The Insider Fire Sale
Call it the millennium rush to cash out—how telecom carrier executives sold out the investors and profited handsomely from the telecom bubble.
Source: Thomson Financial

They were some of the winners in the debacle that saw the CLEC sector's total capitalization drop from $86.5 billion at its peak in 1999 to a total of about $3 billion in early 2001. And this was before the fiascos at WorldCom, Global Crossing, Teligent, and Winstar! (For a complete list of "Mile Nigh Millionaires" see Appendixes A and B.)

A Trip to the Cleaners

The bankruptcy hit parade continued, but Forstmann Little was not ready to throw in the towel on its telecom investments. In April 2001, Teddy's crew invested another $250 million in XO Communications. Just three months later, Moody's, one of the two top credit-rating agencies, cut its rating on both McLeodUSA and XO Communications to junk status, which made borrowing money from the debt markets almost impossible. Such moves are signs of an impending bankruptcy—sort of like being on financial deathwatch.

In May 2001, Forstmann Little invested another $100 million for a 20 percent stake in McLeodUSA. It would continue to support XO and McLeodUSA, even as the telecom malaise continued to crush their revenues and profits. By the time the music stopped, Forstmann Little had invested about $3 billion in the telecom sector. Teddy's faith was repaid with McLeodUSA filing for bankruptcy in January 2002 and XO Communications doing the same in June 2002.

On February 25, 2002, one of Forstmann's limited partners, the State of Connecticut, filed a lawsuit against the once respected fund. "This firm *Enronized* Connecticut through its blatant abuse of trust . . . we want more than the $100 million-plus they wasted and wiped out. We want to make Forstmann Little the poster child for fair-dealing in the investment community," said Connecticut Attorney General Richard Blumenthal.[12] Teddy Forstmann didn't just lose $3 billion. His fling with the telecom sector cost him a reputation built on 25 years of investing experience.

Anatomy of Failure

So why did CLECs in all their variations really fail? There are many conspiracy theories involving the Baby Bells and the way they slowly asphyxiated their upstart competitors. But those are just conspiracy theories. The Baby Bells did whatever they had to do in order to protect their local voice business franchises. Not protecting themselves would have been unfair to their shareholders. The role of the Baby Bells in the death of CLECs is overblown.

Here is why I think the CLEC sector collapsed:

- The Telecommunications Act of 1996 was ill-conceived and was mostly a voice-centric deregulation. The U.S. Federal Communications Commission never really took into account the data (read Internet) side of the business.
- CLECs raised too much money, mostly in debt, and were paying too much interest on the money they raised. Wall Street's bankers, who survive on investment banking fees, had to help them raise money—after all, everyone on Wall Street wants a summer home in East Hampton and a Porsche.

- Easy money prompted companies to build their own nationwide networks at humongous costs, even as the source of demand for those networks was unclear.
- Most CLEC business plans were essentially going after one market—small- and medium-sized businesses. All CLECs, it seemed, had the same business plan.
- Despite its broadband pretensions, the CLEC sector lost its focus early on when it tried to sell voice phone service, thus threatening the regional Bells. The only way CLECs could succeed in the voice market was to compete on price. Had they rolled out true bundled services, they would have had a better chance of winning and retaining market share.
- Instead of fighting the Baby Bells, the CLECs ended up fighting among themselves, in addition to fighting with the Baby Bells!
- The Bells survived because they circled the wagons and used every dirty trick to save themselves from destruction. And there is no business manual that would instruct any differently. The CLECs had underestimated the Baby Bells' survival instincts.
- Selling to someone else, like a WorldCom, was the ultimate exit strategy.
- And lastly, people forget that it took more than 100 years to build the Bell System. No amount of money or deregulation would have allowed a start-up to replicate that network in three years.

9

THE HOUSE (OF CARDS) THAT JACK BUILT

Merrill Lynch's dot-com huckster Henry Blodgett and Morgan Stanley's Mary "Queen of the Net" Meeker may have gotten all the headlines, but it was the behind-the-scenes operator, Jack Grubman, a Salomon Smith Barney telecom analyst, who was the real power broker, the "consigliore," as his colleagues called him. Grubman personified the winner-take-all culture of the nineties. He was in bed with not one but several competing broadbandits. It was as if Grubman was starring in his own version of *The Bachelor*, only he picked all of the bachelorettes, not just one. On the one hand, he was a quasi–personal adviser to WorldCom's Bernie Ebbers, while on the other hand, he helped recruit Qwest's CEO Joe Nacchio.

This modern-day snake oil salesman was adept at playing both sides of the fence and had top telecom executives as well as investment giants like Fidelity and Janus eating out of his hands, with both sides acting upon his advice. All along, though, he was beholden to the investment banking division of his bank, Salomon Smith Barney. Instead of calling these polarities a conflict of interest, he called it a "synergy." According to *BusinessWeek*, since 1996 Salomon helped 81 telecom companies raise $190 billion in debt and equity, earning fees of more than $1.2 billion—more than any other firm on Wall Street—and today both Salomon and Grubman are under investigation for Grubman's ill-fated recommendations.[1]

While some may say that it is unfair to single out Jack Grubman, his role in the broadband bubble cannot be understated. It is difficult to accept that this savvy, intelligent, and once-ethical man, who cohabited with the murky world of broadbandits, didn't know about the shenanigans inside the companies he was hyping. His actions (and sometimes inaction) ensured that the broadbandits stuffed their own coffers and then conveniently cashed out, in what will go down in business annals as the great $750 billion broadband heist.

Grubman focused his energies on top executives like WorldCom CEO Bernie Ebbers, convincing them to borrow billions with the help of Salomon to grow their broadband networks. This, even though no one, least of all Grubman, was sure where the demand for this capacity would come from. Boosterism was like second nature to the man who was a textbook example of the American dream before it became a nightmare on Wall Street. For nearly a decade, whatever Grubman said was deemed the gospel.

■ ■ ■

The Jack Grubman story begins in Oxford Circle, a neighborhood in northeast Philadelphia. In October 1953, Jack Benjamin Grubman was born to Isadore "Izzy" Grubman, a former boxer turned city engineer, and Mildred Grubman, a dressmaker. The Grubmans lived on Magee Street in a modest dwelling. They doted on little Jack, who was born after three miscarriages, and hoped that he would rise above their working class milieu.

Oxford Circle was a blue-collar neighborhood. Most of the Grubmans' neighbors were firemen, cops, or construction workers. The only difference between Jack and the other kids in his neighborhood was that, like his old man Izzy, Jack was a whiz at numbers and quick with his fists.

Izzy, a former Golden Gloves light heavyweight champion, taught Jack how to box. Boxing would become a passion for Jack; he would later decorate his Salomon Smith Barney office with boxing memorabilia, including a signed photo of Muhammad Ali.

After graduating from high school, Jack attended Boston University and received a bachelor's degree in mathematics in 1975. Later, he attended Columbia University, where he received a master's degree in probability theory. This love for numbers helped him snag a job with AT&T as an analyst. A competitive streak and a desire to rise quickly pushed him to work 14 hours a day and then boast about it later. "He was extremely bright, analytical, and hard to supervise," Sam Ginn, a former AT&T executive and later chief executive of the cell phone service provider, Air Touch Communications, told the *Wall Street Journal*.[2]

Grubman wore his working-class stripes with pride. "I didn't grow up in Greenwich. I didn't go to prep school," he later told *BusinessWeek*.[3] (He may not have gone to prep school in Philly, but he did attend two private universities—Boston University, which is not public, contrary to what many think, and Columbia University.) Grubman was arrogant and often confrontational. He caused waves at AT&T a few months after he arrived by figuring out that the math behind the computer model used by AT&T to predict changing phone call prices and their impact on consumer demand was wrong. AT&T stopped using the model, but Jack was on the blacklist. His bare knuckles, take-no-prisoners approach wasn't making him any friends.

Tired of the political machinations at AT&T, Grubman quit Ma Bell in 1985 and joined Paine Webber, a New York brokerage firm, as a telecom analyst. He got off to a rough start at Paine Webber when he flunked his NASD exam, a requirement for all investment professionals. He succeeded on his second try, though. It was around this time that the lies began. He told his employers that he had gone to Massachusetts Institute of Technology and that he grew up in gritty South Philly. "I probably felt insecure," he told *BusinessWeek* later.[4] (It's hard to understand the root of Grubman's insecurities—after all, both BU and Columbia are respectable schools.)

But nobody seemed to notice these half-truths on his resume, dazzled as they were by Grubman's expertise in analyzing the new telecom landscape. In 1984, AT&T had been broken up by federal trustbusters into eight companies: Ma Bell and the seven Baby Bells. There was a dearth of professionals who understood the telecom business as well as Jack did. And he was hard working—he would spend long hours and even weekends at work. His wife, LuAnn, whom he met while the couple worked for AT&T, must have spent a lot of time waiting for him in their apartment on fashionable 12th Street in Manhattan. But all that hard work was paying off: Six years into his tenure at Paine Webber, in 1991, Grubman made $1 million in salary and bonuses.

Grubman's gutsy calls and knowledge of the Bell system also helped him make double-digit profits for his clients. Having befriended Bernie Ebbers, then the chief executive of LDDS, Grubman got a closer look at the independent long-distance business; by 1993, he had helped turn Bernie's company into a growth stock. The word on the street was that Grubman was also advising Ebbers on what companies to acquire. Jack's reputation was spreading and he became a sought-after source for reporters on developments in the telecom industry.

In late 1993, Grubman caught the attention of Eduardo Mestre, head of Salomon Brothers' telecom banking group. The Cuban-born banker had heard good things about Jack, and the two had set up a meeting at the Yale Club near Grand Central Station in New York City. But Grubman apparently got lost on the way to the meeting, and upon arriving, he reportedly yelled at his future boss for providing the wrong address. Mestre's star was ascendant in Salomon Brothers, and he needed Jack. Paine Webber's business was advising retail investors and some institutional clients, but Salomon Brothers was a traders' firm. Its equity research arm was more geared towards large clients. And that suited Jack just fine. Since most Wall Street firms dole out bonuses either around or after Christmas, Grubman waited before announcing his departure. On March 17, 1994, Grubman joined Salomon Brothers, where the high-testosterone culture and aggressive style meshed well with Grubman's own temperament. Mestre and Grubman later became the rainmakers for what became a telecom banking powerhouse.

Grubman's reputation was made in 1995, when he cut his rating on AT&T from a "buy" to a "hold" and reduced his earnings estimate on the company for 1995 and 1996. A "hold" rating is a subtle way of telling investors, "Don't buy this stock." His big call came right ahead of the Telecommunications Act of 1996, and he correctly figured that

AT&T would be one of the companies that could come out a loser when these regulations went into effect. Grubman figured that the act would increase competition for AT&T, which would have little time to hang on to its long distance lead. It was a prescient call. When Grubman downgraded AT&T on April 24, 1995, the stock fell $1.25 to $49 a share. This call pissed off people in Basking Ridge, New Jersey, AT&T's home, to no end. It also shut out Salomon from the forthcoming IPO for the AT&T spin-off, Lucent Technologies—an offering that generated $103 million in investment banking fees for many on Wall Street.

But Jack had cemented his reputation. Money managers loved him. AT&T stock was treading water, and shares of some of the new names Grubman was pushing—MFS Communications, MCI, and LDDS— kept going higher and higher. Grubman had a golden rolodex. Money managers believed that Grubman had the ultimate access to the latest information. After all, he played pool and drank beer with Bernie Ebbers and hobnobbed with other top executives at phone companies. In the process he lined his pockets as well. Since Grubman had no qualms about skirting the line between research and investment banking, his Ebbers connection resulted in Salomon advising LDDS, which later became WorldCom, on more than 65 acquisitions from 1994 on and helped the firm raise around $24 billion in debt, which resulted in fees in excess of $140 million for the New York–based investment bank.

■ ■ ■

The Telecom Act of 1996 brought the dawn of a new era in telecommunications and on Wall Street. Grubman knew a deregulated telecommunications industry would result in new start-ups that would need advice and money. He and Mestre were ready for all the newcomers. In 1996, Grubman's rise to absolute domination of the broadband world began. AT&T executives were quitting in droves to start their own phone companies, and they were all calling Jack. Of course, at the time, there were other influential analysts, like Dan Reingold of Merrill Lynch, who played Roger Ebert to Grubman's Gene Siskel, as the *New York Times* put it. In 1996, Reingold was number one in the annual ranking of telecommunications analysts by *Institutional Investor* magazine, and Grubman was second. But ranking aside, it always seemed that Grubman was the man of the moment.

Grubman also carefully used the media to bolster his image. He was everywhere, easy to reach, had a wry sense of humor, and often had a quip that made for good copy. His close connections with Ebbers made him the guy on the inside, and reporters lapped up his every word. The *New York Times* called him a "swashbuckler, who boasts about his close ties to the chief executives of the big telephone companies and whose research reports sometimes read more like polemics than dispassionate studies."[5]

Grubman's rise to the top was helped by the numerous acquisitions WorldCom was making—acquisitions that Grubman was encouraging WorldCom to make. And of course, Salomon and Grubman stood to benefit from the deals through investment banking fees. In August 1996, when WorldCom made a play for MFS Communications, Grubman jumped to pump up Bernie's ego. "He's (Bernie Ebbers) organically smart. He's very shrewd. He does not believe in management by committee. He trusts his instincts and then has the guts to act on them. Anyone in this industry who dismisses Bernie Ebbers will find him eating their lunch," Jack Grubman told *USA Today*.[6] (When WorldCom eventually bought Jim Crowe's MFS Communications for $14 billion in 1996, Salomon took home a cool $7.5 million in investment banking fees.)

By the end of 1996, it was quite clear that when it came to telecom, all roads led to Grubman. Even high-profile and successful businessmen—some worth billions—were taking his advice. Phil Anschutz, the reclusive billionaire backer of Qwest Communications, was one of them. During the course of a conversation, Anschutz mentioned he was looking for a chief executive to run Qwest, which was then relatively small. "I bet I can get you Joe Nacchio," Jack told Anschutz.[7] And he did. In early 1997, Nacchio quit his job as the head of AT&T's $26 billion-a-year consumer business and went to work for Qwest in Denver.

"Jack has helped me make a lot of money," gushed Jacqueline Cormier of RCM Management, a money management firm based in San Francisco.[8] Other money managers were equally complimentary in describing Grubman and would ring him for advice on many of the new companies that were going public. For Salomon, Grubman was proving to be a growth stock: In 1996, he helped bring in about $60 million in investment banking revenue. In turn, Grubman went home with $3.5 million that year, making him one of the best-paid analysts on Wall Street.

Even the mostly skeptical *Wall Street Journal* took notice. In a March 1997 article, the paper described him as a "swashbuckling deal broker who can sometimes make or break a telecom merger or stock offering." The newspaper also noted that Grubman had no qualms forgoing his independence as a research analyst and tailoring his opinions in order to win investment banking business from corporations, adding, "Grubman is emblematic of a new breed of Wall Street analyst." He was hardly concerned about crossing the ethical line that keeps an analyst separate from the investment banking arm of his firm.

When Qwest went public in March 1997, Salomon was one of the underwriters—after all, Jack had introduced Nacchio to Anschutz. Grubman and Salomon also helped Metromedia Fiber, Teligent, and Nextlink with their IPOs. Not surprisingly, all of these companies were given a buy rating from the get-go. This would become a dangerous pattern for Grubman and Salomon. The red flag should have been raised when Grubman told the *Wall Street Journal*, "It goes without saying that if you do a company's IPO, you are going to have a buy (on the stock), because frankly if you don't you shouldn't be doing the deal."[9] In other words, Grubman was perfectly content to be bullish on Salomon's customers in research reports that were, theoretically, supposed to be unbiased. And he would stay bullish to the very end.

But few paid any attention to this, for America was in the grip of a bull market like never before. Unknown dot-coms were raising billions of dollars from the market, folks like John Sidgmore of UUNet were talking up the demand for Internet bandwidth, and it seemed a new communications revolution was unfolding in front of everyone's eyes. New technologies, the World Wide Web, the Internet, e-mail, and fiber optics made everyone myopic. Investors were looking at the world through rose-colored glasses, where stocks only went up, where markets defied gravity, and where Jack Grubman was as close to a deity as a mere mortal could get.

Grubman's thesis—build it (the network) and they (customers) will come—was just the kind of message a newly deregulated industry wanted to hear. "He walked around like he was a god. And it was perceived in the industry that he was a god," Elliot Dorbian, a former Salomon Brothers broker, told *Money* magazine.[10]

In 1997, Grubman's connections helped Salomon become a powerful force in all things telecom—the investment bank helped 12 companies raise $2.2 billion from the markets in equity and layer it all with $7.4 billion in debt. For all this Salomon earned $120 million, and Grubman's cut came out to about $7 million. (See Table 9.1.)

In October 1997, when WorldCom decided to make a $29.4 billion hostile bid for MCI, Grubman was right by Bernie Ebbers' side. The tough negotiations and the ensuing battle with GTE Corporation and British Telecom for control of Bill McGowan's MCI would prove to be a big payday for Salomon and add another feather to Grubman's cap.

It is rumored that when British Telecom made a cash-and-stock bid for MCI, Salomon tried to exploit the difference between the then MCI stock price and BT's offer. This financial technique was used to take advantage of small price differences between the proposed acquisition price and the then current price, and is used when those placing the bet believe that the deal will proceed without a hitch. While the exact mechanics of the Salomon bet are not known, arbitrage would have cost Salomon millions of dollars if something went wrong.[11] Grubman the expert didn't see this coming—he incorrectly predicted that BT would not renegotiate terms if MCI didn't meet its revenue and earnings target. MCI did issue a profit warning, and British Telecom was still interested, but lowered its offer from $24 billion to $19 billion. Salomon Brothers stood to lose

Table 9.1 Club Jack: Membership Billions Reward Millions
Investment banking relationship with Jack Grubman and Salomon Smith Barney meant huge profits for the decision makers at telecom companies.

Member	Affiliation	Banking Fees (mm)	IPO-related profits (mm)
Phil Anschutz	Qwest	$ 37.5	$ 4.8
Bernie Ebbers	WorldCom	107	11.5
Joe Nacchio	Qwest	37.5	1
Clark McLeod	McLeodUSA	50	9.4
Stephen Garofalo	Metromedia Fiber	47.15	1.5

Source: State of New York and Eliot Spitzer v. Philip Anschutz, Bernie Ebbers, Clark McLeod, Stephen Garofalo, and Joseph Nacchio, Supreme Court of the State of New York, September 30, 2002

$100 million,[12] until WorldCom and Bernie came to Grubman's rescue. WorldCom made a higher offer for MCI, and that helped MCI stock go up and saved Salomon's bet. Grubman got $10 million in 1997 from his ever-grateful bosses.

Later in 1997, when Salomon was acquired by Travelers' Group and merged with Smith Barney's retail brokerage channel, Grubman's star rose even higher. He could literally move the markets. By 1998, it wasn't just the upstarts making a beeline for Grubman's office. Even old industry stalwarts like Grubman's former employer AT&T were calling him to get a feel for how investors would react to certain corporate moves and developments. Former AT&T executives say that it was always, "What does Jack think?"

In February 1998, when the Baby Bell SBC Communications wanted to buy rival Ameritech, SBC's senior executives invited Grubman to a private meeting to seek his advice. SBC executives asked him what they should do in order to grow and stay alive in the newly deregulated environment. Grubman said buy and grow or get acquired—standard advice he gave to all telecom companies. "Look at how fast WorldCom has grown" was the subtle message. Why not do the same? Of course, more mergers and acquisitions meant more investment banking and advisory fees for Grubman and Salomon. In May 1998, SBC followed up with a $72 billion purchase of Ameritech. Salomon was one of the advisers and got over $33 million for its efforts.[13]

"There are others who may model better. There are others who may pick stocks better. [Grubman] knows more about what's going on in the industry than anybody," Rob Gensler, portfolio manager of T. Rowe Price Media & Telecom funds, told *BusinessWeek* later.[14] Grubman's research reports or rating changes were major news for Wall Street. His opinions would be picked up by media outlets such as CNBC, which would feed them to the masses. Investors, both retail and institutional, would rush to buy the stocks based on his opinions—or sell them. The belief was that no one could see the telecom market better than Grubman. Such was Jack's influence that the minute he said a stock was a great buy, the stock would soar heavenward. For instance, in late 1997, when Grubman issued a buy recommendation on Metromedia Fiber Network, an intracity broadband infrastructure provider, the stock grew almost 500 percent over the next 12 months. Similarly, a thumbs-up from Grubman sent Qwest and Global Crossing soaring in 1998.

It was this "expertise" and Grubman's contacts that made the white-shoe investment banking giant Goldman Sachs sit up and take notice. They wanted to get Grubman to do their bidding. Sometime in the spring of 1998, Goldman made a play for Grubman, reportedly offering him $8 million for his first year at Goldman, plus stock that would be worth about $29 million when Goldman went public in 1999. The offer was so high because Goldman thought Jack would bring about $100 to $150 million in business.[15] Goldman was so aggressive in courting that Deryck Maughan, Salomon co–chief executive at the time, called his Goldman counterpart, Jon Corzine (now a New Jersey senator), de-manding to know what was going on, reported the *Wall Street Journal*. Salomon countered Goldman's offer and gave Grubman a $25 million stock and cash package to prevent him from defecting.

It seems everyone wanted Jack!

Welcome to Club Jack

It was the summer of 1998, and Gary Winnick desperately needed a friend. Winnick, the chairman of the telecom upstart Global Crossing, needed someone on Wall Street who understood telecom, was well con-nected, and was powerful enough to raise billions from the market on an as-needed basis. Enter Grubman—the man who became a cheerleader for one and all. The two men became close friends, chatting on the phone on an almost daily basis. When it came time to go public, Grub-man and his Salomon Smith Barney cronies hit the road peddling what would become Global Double-Crossing.

After a successful offering that raised $399 million, Grubman issued his investment thesis on the company: "Global Crossing is building a truly unique and valuable asset." In February 1999, when Winnick was looking for a new chief executive, Grubman brought him Robert An-nunziata, who had sold his Staten Island–based upstart Teleport to AT&T for $11.3 billion. In March 1999, when Global Crossing tried to buy Frontier Communications, Salomon and Grubman were on hand. Hell, Grubman was such a nice guy that he even made a presentation to Global Crossing's board of directors, convincing them that the $11.2 bil-lion bid for Frontier was a good one. (Global Crossing ultimately bought Frontier for around $8 billion.)

And when Winnick's crew decided that they wanted to go after US West, a Denver-based Baby Bell, Grubman and company were at their beck and call. Later, when the deal fell apart, US West had to buy 9.5 percent of Global's shares as a penalty, and Jack advised Gary to sell $350 million worth of his stock. Winnick wanted to sell more, but Jack advised him against it, as it would make large investors nervous about the prospects of the company.[16] No wonder Winnick later described Grubman as the "Bruce Springsteen of Telecom." In 1998 and 1999, Winnick and Grubman were in close contact, but that relationship came to an end in 2000, when Robert Annunziata left Global Crossing to work for PF.Net, a competitor, and Leo Hindery joined as Global Crossing's CEO. Grubman didn't like Hindery, and apparently the feeling was mutual.

Winnick wasn't the only one who swooned when Grubman came into a room. In an interview with *Business Week*, Allegiance Telecom chief executive Royce Holland recalled that during a visit to Boston in 1998, he accompanied Grubman to a meeting where money managers were present. "It was like the messiah had come to town," Holland told *Business-Week*.[17] "He's almost a demigod—he's the king of telecom," gushed Robert Knowling, chief executive of Covad Communications, in that same article.

Another man who thought the world of Grubman was William Rouhana, the founder and chief executive of the fixed wireless carrier Winstar Communications. Grubman helped Winstar do a $650 million private placement in March 1998. Salomon and CSFB underwrote that deal. Salomon would help raise even more money for the fixed wireless company in the months to come. And whenever it seemed investors were tiring of fixed wireless, Grubman would pump up the stock. Through much of 1999, Grubman had the Midas touch. Salomon Smith Barney raked in about $24 million in investment banking fees, and the money managers who listened to Grubman were looking awfully smart, especially when they went on CNBC, the official cable network of the 1990s bull market.

While membership in Club Jack was expensive—millions in investment banking fees—the returns were equally high for some senior executives. To reward executives for their business, Salomon gave shares of some of their hot telecom initial public offerings to their favorite executives. In turn, these executives could flip the shares on the first day of trading, thus netting millions for minimal work. This practice of spinning was quite widespread on Wall Street. Credit Suisse First Boston and Goldman Sachs were two other investment powerhouses that have been accused of the stock version of "commercial bribery."[18]

Between 1996 and August 2000, Bernie "Grubman's buddy" Ebbers received 21 hot IPO allocations, including those of Williams Communications, Juno Online, and Rhythms NetConnections. His net proceeds were about $11.5 million.[19] From 1997 to 2001, Qwest used Salomon on 18 different deals and paid the bank about $38 million. Qwest backer Phil Anschutz received 57 IPO allocations over that period and profited to the tune of $4.8 million. His lieutenant, "Jumping" Joe Nacchio, got stock in 42 IPOs and sold them for a profit of over $1 million. Metromedia Fiber Network (MFN) paid Salomon $48 million for work on 15 deals, and MFN founder Stephen Garofalo received shares in 37 IPOs, profiting to the tune of $1.5 million. But even that was chump change compared to the $9.4 million that Clark McLeod, founder of Cedar Rapids, Iowa–based McLeod, a next-generation carrier, made from allocations in 32 hot IPOs.[20]

No wonder everyone wanted to be part of Jack's inner circle. In fact, some at WorldCom were angry that Bernie was getting all the gravy and complained bitterly that they were not getting enough IPO shares. Scott Sullivan, WorldCom's CFO, was one of those who protested most aggressively for not getting enough.

These conflicts of interest on the part of both Grubman and the chief executives were an offshoot of the greed culture that permeated Wall Street and corporate America at the time. In the 1990s, the IPO market was like heroin for the new era of greed. Jack was the dealer, and the corporate titans the addicts. But Jack was merely a symbol of the crumbling ethical value system. Morals were sacrificed in an attempt to make easy money. The losers were small investors, who hadn't a hope in this rigged game.

Joker or King

By the fall of 2000, things were slowly getting out of Grubman's control. Published reports indicate that he was working 15-hour days, traveling all the time, and had no time to spend with his twins and wife LuAnn, who now lived in a $6.2 million townhouse on Manhattan's Upper East

Side. Grubman was the subject of a flattering *Business Week* profile, but his troubles at Salomon were just beginning to appear.

According to Salomon's internal documents, Grubman brought in about $255 million in investment banking revenues in 1998, and the number increased to $359 million in 1999. But in 2000 it dipped to $331 million, reflecting the downturn in the stock market that year. These are still staggering numbers by all accounts, but the telecom nuclear winter was on the horizon. The endless chain of telecom bankruptcies was beginning, and the stocks Grubman had been touting were being pounded on Wall Street. In November 2000, Global Crossing was down to $16 a share, while Qwest slid to $37.75 a share (on November 30, 2000) from an all-time high of $64 a share on March 2, 2000. Metromedia Fiber was trading at around $12 a share in November 2000, and WorldCom, the Teflon telecom stock, was at $16 a share. It's hard to digest that Grubman, who took pride in his rolodex and contacts with executives at most broadband companies, didn't know the market reality. He had a buy rating on most of his stocks.

Grubman would say "Buy," Salomon Smith Barney's retail brokers would push those stocks to millions of clients, and the stocks—be they WorldCom, Global Crossing, or Metromedia Fiber—would go up. This would make the chief executives very happy, and for the mutual funds and hedge funds, Grubman remained the messiah of moola. The little guys who were buying on Grubman's advice would feel happy and count their paper profits. But when things started to go wrong, Grubman realized that he couldn't get off the treadmill of greed, and this was costing a lot of Salomon Smith Barney's retail clients a lot of money. These retail investors were angry, and they directed their venom at Grubman.

As a result, after a time, Salomon's own brokers were reluctant to tout anything Grubman recommended. "Jack Grubman is not an analyst, he is an investment banker," lamented one retail broker.[21] Another Salomon broker was even more petulant when he suggested, "On the retail side the damage he has done is a disgrace. I hope many clients sue."[22] It seemed like Grubman was losing traction with the very people who had helped him keep broadband shares moving up like a helium balloon.

"Grubman is an investment bank whore! When is the firm going to stop pimping him?"[23] asked one broker, whose clients had lost thousands of dollars buying stocks like WorldCom and holding them in their accounts for long after things started going wrong.[24] From 1994 to 1999, he might have been king, but in the new millennium he had become a puppet of the investment bankers, a joker in the pack.

In the early days it was Jack who was bringing deals to Salomon and dictating terms, but as the markets tanked and Jack lost some of his influence on the investors, the bankers pushed him to compromise even more. Documents show that he was increasingly becoming a puppet in the hands of Salomon's investment banking group. And nowhere was it more obvious than in 2001, when Grubman lost all his credibility.

As the New Year began, the debris from the burst broadband bubble was catching Grubman in the face. His bosses were upset because revenues from telecom-related investment banking activities were down to $166 million, half of what Grubman had managed to generate in 2000. Grubman was still working hard, but bankruptcies were starting as stocks tanked. On top of that, Grubman was losing any influence he had over the large institutional investors.

One stock that tarnished Grubman's reputation with large mutual fund clients was that of Winstar Communications, the fixed wireless operator that had been a long-standing client of Salomon Brothers. The stock was trading at about $11 a share at the start of 2001. Then on January 25, 2001, Grubman released a research report on the company that reiterated his buy rating and set a $50 price target. At the time, Winstar, which had paid Salomon Smith Barney about $24 million in exchange for raising $5.6 billion over the three previous years, was in a perilous state. For the year ending December 31, 2000, the company had lost about $894 million on sales of about $759.3 million.

Grubman, who had started following Winstar in January 1998, never changed his buy rating on the stock. However, by January 2001, most investors were wise to the ways of Winstar. And no one really believed it when Grubman said to buy this stock. Apparently, even Grubman didn't believe it. On February 20, 2001, in an e-mail to an institutional client, Grubman's assistant Christine Gochuico hinted that he should sell the stock in the low-$20s. At the time, Winstar was trading at $13 a share. Three days later, on February 23, 2001, Gochuico received another query about Winstar's viability, and this time she was even more candid when she replied via e-mail, "Hope it doesn't go to zero since we have been so vocal on it."[25] The $50 dollar price target Grubman was touting was quite bogus.

Manuel Asensio, a well-known short-seller, believed that Winstar was a sham and called Grubman's bluff in a press release dated March 8, 2001. "Winstar owes over $1.3 billion to banks, mostly incurred to repay prior existing vendor debt, and has another $1.6 billion of high-yield bond debt. These debts alone are problematic. Winstar has already given the banks substantially all of its current and future assets as collateral. In addition to this $2.9 billion of troubled debt, Winstar has another $1.6 billion of vendor debt and other liabilities," the press release said.

Winstar was getting pummeled, and Grubman's staff was getting e-mails and phone calls from institutional clients, concerned about their investments in the stock. Grubman's credibility was being shredded in public. Sherlyn McMahon, a senior analyst associate in Grubman's group, wrote him an e-mail expressing a client's concern about the stocks the telecom group was recommending. "She [the investor] thinks that we make ourselves look stupid by recommending names right up to bank-ruptcy, like WCII [Winstar], XOXO [XO Communications], MFNX [Metromedia Fiber Networks], etc. She understands the investment banking aspect," McMahon wrote.[26]

These stocks were plummeting faster than an asteroid headed to-wards Earth. Grubman, it seems knew that things were bad, but he didn't say anything publicly. Privately, he grumbled! Grubman ex-pressed concern that the investment bank was improperly pressuring the research group to continue issuing positive reports on clients. He wrote an e-mail to Kevin McCaffrey, head of Salomon Smith Barney's director of research (and Jack's boss) saying, "[M]ost of our banking clients are going to zero and you know I wanted to downgrade them months ago but got huge pushback from banking. I wonder what use bankers are if all they can depend on to get business is analysts who rec-ommend their banking clients."[27] It was a telling e-mail that high-lighted the role of Salomon's investment bankers in creating the broadband bubble of the 1990s.

For Grubman, things came to a head on February 21, 2001, when he issued a bullish report on Focal Communications, a Johnny-come-lately phone company with little or no prospects. Up until that point, Focal had paid Salomon Smith Barney about $10 million in investment banking fees for three transactions. The company's stock was trading at $15.50 a share, and Grubman had predictably accorded the stock a buy rating.

During the course of the day, he learned that Focal was unhappy about some of the content of the report and had bitterly complained about it.

In an e-mail, an institutional investor asked Grubman's associate, McMahon, "Focal and McLeod [McLeodUSA] are pigs aren't they?" Grubman simply lost it, and perhaps in a flash of conscience, he wrote an e-mail to two Salomon bankers: "[If] I so much as hear one more f——— [his dashes] peep out of them [Focal] we will put the proper rating (i.e., 4 not even 3) on this stock which every single smart buy sider [institutional investor] feels is going to zero. We lost credibility on MCLD and XO because we support pigs like Focal." (According to Salomon's ratings system, 4 means underperform and 3 means neutral.)

On April 18, 2001, Winstar went bankrupt. It wasn't an isolated case. Grubman's record was like that of a serial killer—he left dead companies in the wake of his buy recommendations. In an e-mail to his assistant Gochuico, he admitted not doing enough due diligence on Winstar. "If anything, the record shows we support our banking clients too well for too long," he wrote in an e-mail to Christine.

Worried about his reputation (a bit late in the game for that) he tried to change his rating on Focal again on April 18, 2001, but the bankers pushed back. By August 13, 2001, when Focal was trading at $1.24 a share, Grubman finally downgraded the stock. It was enough for the bankers—Salomon had raked in about $12 million in investment banking fees between February 21 and August 13. The return on keeping a buy rating: $2 million a month!

But these were minor problems compared to two pending disasters, Global Crossing and WorldCom. Having hit a peak of $61.81 in February 2000, Global Crossing was slowly sliding downwards, but that didn't worry Grubman. "These are historic opportunities to buy world-class assets such as Global Crossing that are evolving into world-class operating businesses at compelling value," Grubman wrote in a report dated June 18, 2001. The stock was trading at $7.68 a share, down 88 percent from its peak. It would sink to $1.90 a share by October 1, 2001. "The bottom line is we believe that Global Crossing is not a potential bankruptcy candidate in the near term," Grubman wrote that day. Four months later, Global Crossing filed for bankruptcy.

Then came the ultimate shocker: On February 8, 2002, WorldCom cut its revenue and earnings projections for 2002 and said it would take a

charge of $15 to $20 billion to write down the value of some acquired operations. Earnings in the fourth quarter were off 64 percent. As goes World-
Com, so goes Grubman. The bad news from both Global and WorldCom
was like a left-right punch to Grubman's face. He was on the mat, bleeding,
but he still had some fight left in him. As the number of bankruptcies increased in the spring of 2002 the media spotlight on Grubman was glar-
ingly harsh. A defiant Grubman defended himself in a *Money* magazine
story: "If you took the emerging telecom names in total from their peaks in
March 2000 to today, there was a total of $230 billion or so of market cap
loss. Do you know during that time Cisco lost almost $450 billion of mar-
ket cap by itself?"[28] Of course, if Grubman hadn't hyped those telecom
stocks, it wouldn't have created a broadband equipment bubble, and companies like Cisco wouldn't have seen their stocks run up to irrational levels.

On April 30, 2002, Bernie Ebbers was fired by his own board of directors. Investors, who had by now stopped paying heed to Grubman,
were bailing from WorldCom, the company that had essentially made
Grubman. WorldCom's debt was going to be downgraded to junk status, and it was only a matter of time before WorldCom knocked on the
door of Chapter 11, which it did in June 2002. Grubman did his job
and downgraded the stock, as expected, a few days before the bankruptcy. The erstwhile long distance discount reseller had the dubious
distinction of being the biggest bankruptcy in the history of America,
even bigger than Enron. But more than that, the company would admit
that it committed an accounting fraud, which at last count misstated
revenues by about $9 billion.

WorldCom's downfall was a blow from which even Grubman couldn't
recover. For almost 15 years, he had carefully cultivated and nurtured his
relationship with Bernie Ebbers. WorldCom was the kind of client that
investment banks dream of—acquisition-hungry and always looking to
raise more money from the capital markets. Between 1997 and 2001, Salomon Smith Barney got about $107 million in investment banking fees
from WorldCom. If this meant steering some hot IPO shares worth millions to Ebbers, then so be it. If Grubman had to share his revenue models with Sullivan or prep the company about the questions he would ask
on a conference call, those were small compromises.

The story goes that once during a WorldCom quarterly conference
call, Bernie and Grubman discussed the golf game they had played a

week earlier. It was Grubman's way of flexing his muscles and letting the rest of the world know who was the ax on WorldCom. "Ax" is an insider term for an analyst who can make or break a stock with thumbs up or down. WorldCom, through Ebbers, was attached at the hip to Salomon Smith Barney, through Grubman. Salomon's parent company, Citigroup, had loaned Ebbers $679 million. It's speculated that Grubman got a finder's fee—about 10 to 15 percent of the profits made by the Citigroup on the loan—for arranging this deal.[29]

The ensuing hullabaloo around WorldCom was too much for Grubman. He was hauled up in front of the House Committee on Financial Affairs that was investigating the WorldCom scandal. He was grilled mercilessly. Signs of stress were showing on his face as he faced angry politicians on July 8, 2002. "WorldCom is a company that I believed in wholeheartedly for a long time. It fit my long-held, honestly held investment thesis that the newer, more nimbler companies would create value," he said.[30]

During the hearings, not once did he come out and say that World-Com had conned him. He defended his employer and his friend Bernie to the very end. But he was clearly becoming a liability. Salomon Smith Barney had to cut him loose. After all, a month earlier in Washington, D.C., during Congressional hearings, Grubman had said, "First and foremost, your value and worth as an analyst to the firm you work for and to banking clients starts and stops with your credibility in the marketplace with investors. And if you blow that, then you have no value to anyone."[31] He had helped his firm generate billions in business, whether it came from investment banking, underwriting or stock trading. It was clear that his gravy train had come to the end of the line.

On August 18, 2002, like Elvis, Grubman left the building. He left with $32 million in cash and stock compensation and a promise from his bosses that they would pay his legal fees. He was being sued by investors who had lost money; the New York State Attorney General Eliot Spitzer and the National Association of Securities Dealers were investigating him; and more legal troubles were sure to follow. In his resignation letter, Grubman wrote, "The relentless series of negative statements about my work, all of which I believe unfairly single me out, has begun to undermine my efforts to analyze telecommunications companies."

He did have a point about other analysts who were all desperate to be mini-Grubmans. He also had a point about the investment bankers at Salomon who pushed Grubman and funded companies they knew had no chance. It is hard to pin all the blame on one guy. What about Salomon's all-star telecom banking team—guys like Tony Whittemore and David Diwik, who left Salomon in November 2000 to work for Deutsche Bank; Tom Jones and Christopher Lawrence, ex-Salomon bankers now gainfully employed at Credit Suisse First Boston; and those who are still at Salomon: John Otto, Eduardo Mestre, and Thomas King? And it is hard to digest that the Citigroup head honcho Sandy Weill didn't have a clue about what was going on. After all, you don't loan the CEO of one of your clients $499 million without checking with the boss.

Grubman was nothing but a used car salesman in a Brooks Brothers suit, a huckster who had schlock to sell. But what about the mutual fund managers and others who bought the stock? They didn't have to listen to Grubman. They all were as much a part of telecom companies' rise and fall and were equally guilty of succumbing to greed. "Some people may not like this because you have to look beyond the sell side analysts and you have to go through the entire supply chain of who buys and sells stock. I believe that over the past certainly half a decade that the entire market has become much more short-term oriented than long-term oriented," said Grubman.[32] "It is the mutual funds and pension funds and money managers out there who increasingly by their clients are getting graded every quarter. Pressure to perform quarter in and quarter out doesn't stop and start with Wall Street. It goes all the way through the supply chain of who manages money and each client at each turn of the corner puts increasing pressure to perform on a quarterly basis. So it is a big issue."

And if that is not enough, now Grubman is involved in a new scandal—Nursery Gate, as the tabloids are calling it, insinuates that he changed his ratings on AT&T in order to please his boss, Sandy Weill, head honcho of Citibank, so that Weill would get Jack's twins into an elite Manhattan nursery school. He is also said to be cooperating with New York Attorney General Eliot Spitzer—and that could lead to bigger fish being caught in the net.

PART III

THE LIGHT
KNIGHTS

10

CANADIAN RHAPSODY

John Roth was frustrated. He had been recently named the chief executive of Nortel, the second largest telephone equipment maker in the world, but he could not find a new glove compartment liner for his Jaguar. An avid car enthusiast, he owned six high-end vehicles, including a Ferrari, a Prowler, and a '67 Corvette. But it was a 1966 E-Type Jaguar that was Roth's pride and joy, and the spare part he sought was nowhere to be found anywhere in Canada. In the summer of 1997, he turned to the Web.

"I started poking around, and within five minutes I'm in a little garage north of London, England—a four-man operation—and he's got my parts. How would I have ever have found this person if he wasn't on the Web? The idea that some guy north of London can carry out a business transaction with someone he's never met in Canada—it struck me what a powerful force this was, and how could something like this ever be stopped?" he later told *MacLean's* magazine.[1] Roth became somewhat of a Web shopaholic. He ended up buying the glove compartment liner, along with a jukebox and a pool table. The process also convinced him that the future of Nortel depended on the Internet.

In 1997, Roth launched a strategy that took the once-sleepy Canadian company on a roller-coaster ride that can only be matched by the Nitro ride at Six Flags Great Adventure Park in New Jersey. Nitro, with a mile-long track, blasts off and rises to 230 feet before plunging 215 feet back to earth at speeds approaching 80 miles per hour. The ride lasts four minutes, but its aftereffects keep you popping Dramamine for weeks.

When Roth was named Nortel CEO, he figured this was his chance to take on AT&T's hardware arm, Lucent, and Cisco Systems in what would become a high-stakes game.

He decided it was time to refocus Nortel's engineering energies on data networking rather than on the traditional world of voice telephony. This was quite a turnaround. Like an aging diva, Roth's Nortel went in for a tummy tuck and a facelift, and, as temporary solutions go, the results were rather alluring! Between 1997 and early 2001, the company spent about $32 billion—much of it in stock—to acquire 19 companies. The more companies Nortel bought, the higher its stock valuation rose; its market capitalization hit a whopping $323.4 billion on July 25, 2000, making Nortel the eighth biggest company in the world, measured by market capitalization.

Not anymore. More than 60,000 Nortel employees have been given the pink slip since February 2001, and at the end of November 2002, the stock was trading at $1.52 a share, down 85 percent from $13.16 a share on October 1, 1997, when Roth took over as chief executive.

■ ■ ■

Born on October 6, 1942, in Lethbridge, Alberta, John Roth was the son of an airline radio operator, which explains his interest in all things wireless. Roth went to engineering school at the prestigious McGill University and graduated in 1966. He spent the next three years at RCA Canada, the Canadian division of the American consumer electronics company, before joining Nortel—then called Bell Northern Telecom—as a design engineer for the company's satellite research division in 1969. He rose steadily through the ranks, happy to be the complete company guy. By the time Roth was 36, in 1978, he'd been named vice president of manufacturing operations. In 1982 he became the president of Bell Northern Telecom, the youngest president in the history of the company, which had been founded in 1895 as Northern Electric.

Despite his growing stature inside Nortel, Roth and his wife Margaret have lived on a 50-acre farm in Caldeon, Ontario, a rural community close to Nortel's headquarters, for a long time. Roth, who is six feet four inches tall, has been described as a quiet, extremely private family man who would rather wear a lumber jacket than a suit. His mannerisms are precise, and he speaks in a deep, steady voice.

When Roth joined Bell Northern Telecom, it was a wholly owned subsidiary of Bell Canada, Canada's premier phone company. In 1971, the company created Bell Northern Telecom as a research and development arm modeled after Bell Labs, owned by AT&T. In 1972, it was sold to the public. By then its name had been shortened to Northern Telecom. After the IPO, Bell Canada held a 90 percent stake in the company and was one of its primary customers. In 1999, Bell Canada, which had been slowly cutting its stake in Northern Telecom, decided to distribute its holdings in the equipment maker to Bell Canada shareholders.

This made Nortel one of the most widely held stocks in Canada and also one of the largest enterprises in the country. It was this behemoth Roth would one day lead as its chief executive. Normally, engineers in technology companies prefer to be left alone, and have no desire to get involved with the management. But, despite being the ringleader of the pocket-protector set inside Northern Telecom, Roth was an ambitious guy. In the early 1990s, when the cellular boom was about to get under way, Roth convinced the top brass at the company to diversify and develop equipment for this fast-growing market. At the time, cellular networks were going from analog to digital, a shift similar to the transition from vinyl records to compact discs, and market demand for digital wireless gear was set to explode. This prescient move put Roth on a fast track to the top.

Roth had a persuasive but understated style that won him many fans at the company. The cellular business proved to be highly lucrative, and any other executive would have taken this success and gone elsewhere. But Roth figured the senior management would do right by him. They did: In 1993, Roth was named the president of wireless networking. In 1994, he convinced Northern Telecom to throw a lot of resources into researching fiber optics and related technologies. These technologies had been in development in Nortel's labs since 1989.

Fiber optics technology transmits information—video, voice, or data—carried on light waves through hair-thin strands of glass, also known as fiber. Each wavelength or color in a light beam can be used to carry huge volumes of information. At the time, the fiber optics business was the preserve of a handful of established players like AT&T's hardware division (now Lucent Technologies) and industrial conglomerates like Siemens, Fujitsu, NEC Corporation, and Pirelli. The demand was predictable, as only a few dozen phone companies built networks to carry voice traffic.

Roth had had his data networking epiphany—he knew the hyperdemand for bandwidth was coming. He realized that the phenomenal growth would require a whole new different class of equipment to transmit data between cities and countries over the Internet. Roth told his bosses that voice traffic—previously the primary source of the company's business—would become a fourth of the total traffic carried by various networks, and that existing networks would soon become clogged because of data traffic's growth. This was a full three years before the dotcom boom, and in many ways it was Roth's foresight that put the company on the fast track.

In 1995, after he was named the chief operating officer of the company, Roth challenged his engineers to develop a system that was four times faster than the optical networking equipment that was being sold mostly by Lucent and a handful of European companies like Marconi and Pirelli. Industry watchers scoffed at the idea. The conventional wisdom at the time was to use a system that transmitted 2.5 gigabits per second in combination with a new technology called Dense Wavelength Division Multiplexing (DWDM), a way to chop light beams into individual colors and then use the different colors to cram more data on the same fiber. Lucent was taking this approach, as was Ciena, which was on the cutting edge of the DWDM technology.

Critics said that the lasers that powered this high-speed, high-capacity fiber-optical transmission system, also known as the OC-192 system, would melt the glass and the laser-generating light. Also, they said, since the glass and light were hard to control under any normal circumstances, the signal would disperse, making it difficult to carry any kind of traffic. Lucent's and Ciena's systems were like a highway with 16 lanes where you could only drive at 50 miles an hour; Nortel, on the other hand, proposed a 4-lane autobahn on which you could travel at 200 miles an hour. Lucent executives, including future chief executive Richard McGinn, openly ridiculed Nortel for developing the system.

But Nortel proved its detractors wrong when, in 1996, it launched its first 10-gigabit per second fiber-optic system, which was four times faster than currently available technology. If all Americans started talking to each other on the phone simultaneously, the traffic of their chatter would fill a tenth of the system's capacity.

Everyone, including Microsoft's Bill Gates, had underestimated the Internet's development and the growing popularity of e-mail. No one figured that buying books from Amazon.com and selling knickknacks on EBay would become national pastimes. Back then most people failed to realize the revolution Netscape's browser would unleash. Instant news, online shopping, and auctions drew tens of thousands of Americans to the Internet. The networks were getting clogged. Roth had been right!

In 1997, when Nortel's then chief executive Jean Monty left to become the chief executive of BCE, Canada's largest phone company, Roth's ship finally came in. Developing the OC-192 system was only half the battle. Traditional phone companies such as AT&T and its progeny, like Ameritech, Pacific Bell, NYNEX, and Bell Atlantic, refused to buy products from Nortel. For decades, these guys had bought equipment from Lucent, so why change now? Worried about recouping the massive investments in developing the technology, Roth hit the road to drum up business.

Figuring that traditional phone companies were less likely to buy his wares, Roth began targeting next-generation carriers like Qwest, Global Crossing, and Level 3 Communications, which weren't attached to buying Lucent's gear and needed equipment to help them build networks with faster speeds and more capacity. The business model for companies like Qwest was to build really big pipes, fill them up with data, and then build more.

First Roth and his boys found the going tough even among the new carriers. Bijan Khosravi, who was one of the people who worked on the OC-192 product, recalled that the company was having a tough time selling it. Qwest, which was in the market for gear, told Nortel no thanks; Qwest CEO Joe Nacchio was going to work with his old pals at AT&T's Lucent. It seemed that five years of development were going down the drain. "We were worried, and we did figure it was all because of the wrong marketing approach," recalled Khosravi, who is now chief executive of Atlanta-based Movaz Networks. "We basically went to

Qwest and made them change their mind, and ultimately that Qwest win changed everything for us."

After that, Nortel started marketing its products quite heavily and soon enough, Roth always found a receptive audience among the likes of Qwest's Joe Nacchio and Level 3's Jim Crowe. But still there was a price to pay.

In 1997, desperate to make a sale, Roth made Nacchio a deal: Pay Nortel up front for a 2.5 gigabit-system, and pay for the rest of the systems when Qwest used the entire capacity. This practice would eventually mutate into a sleight-of-hand practice called "vendor financing" that would become a catalyst in the near demise of Nortel and Lucent. A few months later, as 1997 came to a close, Nacchio would be on the phone with Roth again, looking for more equipment.

Dot Goes Roth.com

It was time to go public with the Nortel story. On April 22, 1998, Roth announced his sweeping new vision of a world in which a large data network would replace the traditional voice networks, and everything from computers to phones to laundry machines would connect to this new lifeline. "Given that data traffic is growing ten times faster than voice traffic, by the turn of the century, we'll be looking at networks here in North America where the traffic will typically be 75 percent data, 25 percent voice," Roth proclaimed.

It was a big boast from a company that at the time had little or no presence, or expertise, in the data networking business and the underlying technologies that made the Internet work. It might have been an expert in optical and voice technologies, but it still lagged in the Internet-based networking technologies. The core technology of the Internet, called Internet Protocol, or IP, was not in fashion at the time, as the phone industry was focused on a more reliable but more expensive technology called Asynchronous Transfer Mode, or ATM. Think of ATM as the Federal Express of networking protocols, and IP as the U.S. Postal Service. At the time, only Cisco Systems had the expertise to make IP equipment. But the corporations loved IP-based products—they were cheaper and easier, as shown by rising sales of Cisco.

It is no surprise that Roth, a quintessential engineer, fell in love with

IP technology, defying the then current wisdom. "I'm convinced the future success of Nortel will depend, to a large extent, on our ability to do for IP networks what we've done for voice networks," Roth wrote in a letter to employees two months after becoming CEO in October 1997. "IP networking will become one of the corporation's core competencies." His grandiose statements were laughable, for at the time, Northern Telecom was a dowdy company that, despite its early success in the optical markets, was mostly known as a maker of PBX systems and phones.

Known for pushing his sports cars to the limit on a track near his home, he had similar tactics in business. He decided to jump headfirst into the IP world. In June 1998, Roth paid $13.4 billion for Bay Networks, a router maker that was, at the time, Cisco Systems' primary competitor.

Formed after the merger of Synoptics and Wellfleet, two networking equipment makers, Bay Networks was lagging behind Cisco on all fronts. Despite good engineering resources, the company was outhustled by Cisco Systems' crackerjack sales force. For Bay, Nortel's merger proposal was heaven-sent, and its acceptance of this proposal was an admission that it could never compete with Cisco. Dave House, Bay Networks' CEO, realized that selling the company would help recover the most value for the shareholders. The original Synoptics-Wellfleet merger had gone horribly wrong, people were leaving the company, and the company's managerial chaos was an open secret. When Roth showed up with a $13.4 billion offer, Bay executives must have said to themselves: "Where do we sign?" "While I wasn't necessarily a big [Dave] House fan overall, I think this was the right move given the circumstances," said one former Bay executive. Bay had essentially misplayed the enterprise market for years. Bay was fundamentally a technology-driven company, not a market-driven one. "The problem was that Bay had misexecuted for so long and had lost so much key talent that it was wounded beyond repair."

Many questioned the wisdom of the deal, saying it was too expensive. Nortel's stock sank 53 percent within four months of the acquisition. The market was once again proven right, for Nortel never really became a force to reckon with in the IP space. "We were a company that moved at the pace of the telephone, and we had to move at the pace of the Web," Roth said in justifying his purchase to *USA Today*.[2] One couldn't blame Roth for thinking IP was the way to go and for buying Bay Networks. It was a telecom equipment supplier that was facing a migration from traditional technologies to IP. Cisco was touting IP's suitability for all kinds of data transport, claiming that the protocol would one day be the universal communications network. Despite its optical expertise, Nortel had been unable to move fast enough or shed its image as a traditional, old-school telecom supplier. The acquisition, Roth thought, would bring in experts who would help the company get IP expertise. (See Table 10.1.)

How Bay Got Bungled

Right after the merger, Roth decided that the name "Northern Telecom" wasn't good enough. For decades a boring old telephone equipment

Table 10.1   Northern Lights

Nortel became an optical powerhouse on its own, but in a bid for global domination went on a massive shopping spree.

Announced Date	Target	Size ($mm)
May 1996	Micom Systems Inc.	$ 150
April 1998	Aptis Communications, Inc.	305
June 1998	Bay Networks, Inc.	9,100
December 1998	Cambrian Systems	300
April 1999	Shasta Networks, Inc.	340
May 1999	X-CEL Communications	n/a
November 1999	Periphonics Corporation	440
January 2000	Qtera Corporation	3,250
February 2000	Dimension Enterprises, Inc.	65
March 2000	Clarify, Inc.	2,100
March 2000	Nortel Networks Broadband Access Inc.	778
March 2000	Xros, Inc.	3,250
April 2000	CoreTek, Inc.	1,430
May 2000	Photonic Technologies	36
July 2000	Architel Systems Corporation	395
July 2000	Alteon WebSystems, Inc.	7,800
August 2000	Sonoma Systems	540
November 2000	Nortel Networks High-Speed Networking Card Unit	110
Total	**18 targets**	**$30,389**

Source: Capital IQ

maker, it was now a superpower in Internet infrastructure and needed a new name to reflect that. In 1998, Northern Telecom was renamed Nortel Networks, in an effort to give it new-economy cachet. Justifying the name change required a press release, and Nortel issued plenty. "The name change reinforces that we're a company that delivers a broad and growing portfolio of integrated network solutions spanning data and telephony," Roth bragged in a press release. "The use of the Nortel Networks brand name better defines for our customers, employees, and investors our market position in the new era of networking being shaped by the Internet revolution. It better defines us as a network solutions company right at the heart of the Internet revolution. You could say that telecom is the 'killer application' of the Internet." All the boasting didn't help.

In 1999, the company was still perceived as an also-ran, largely because Cisco Systems of San Jose was proving to be an unmatchable competitor with better marketing strategy. Nortel decided to take a page out of Cisco's strategy and launched a carefully orchestrated ad campaign. On October 23, 1999, right during the World Series, an ad blitz was launched. "Nortel Networks is building the new high-performance Internet . . . so tell us, what do you want it to be?" the spots said. More ads followed, and this time they featured celebrities such as pop star Elton John, Dallas Cowboys owner Jerry Jones, fashion designer Oscar de la Renta, and musician Carlos Santana.

Despite Nortel's new-economy company aspirations, its culture was essentially old-economy—bureaucratic, slow-moving, and methodical. "This is a company which was used to selling to about a dozen-odd established phone companies," said an insider. Before the Bay acquisition, Nortel had been focused on building voice products precisely engineered to customers' needs and then selling that equipment to these dozen customers. But the Bay Networks acquisition changed the company's engineering focus.

It also brought a stock option culture to Nortel. Instead of getting hefty cash packages, tech companies would dole out options, which gave employees the right to buy company stock at a predetermined price. It was like poison gas for a company that had led a frugal life. When asked about the impact of the Bay acquisition, Roth told the *Wall Street Journal* that "stock options were probably one of the most significant changes we made."

"At first, many of the Bay people were fearful that Nortel would send in a completely new management team that knew nothing about Bay's technology or products and they would cause a huge exodus of people and consequent ruin. This didn't happen. Nortel let Bay operate by itself for quite some time," said Dave Roberts, who was a senior marketing executive at Bay Networks and is now a co-founder of Inkra Networks, a Silicon Valley start-up. "High-level Bay execs were incentivized to stay on. Nortel was great at reopening large accounts that had been closed to Bay. We were able to compete more as peers with Cisco."

But then the problems started. Despite all its IP pretensions, Nortel couldn't get away from its phone company roots. Prior to the Bay acquisition, Nortel could count most of its major accounts on two hands. Most of its customers were phone companies and large network providers, but after the Bay acquisition the company had to deal with a large number of corporate customers. After the acquisition, the newly merged company had to figure out how to deal with literally thousands of customers. "This was a mental shift that the execs just couldn't make," said a former executive with the company.

Despite all these different problems lurking in the background, the main reason Nortel bungled the Bay Networks purchase was fiber optics. Inside the company, the fiber optics division was the top of the heap. The demand from new-generation carriers was so high that fiber-optic gear was literally flying out of the door. Nortel was making boatloads of money from its new optical networking business.

The company organized itself along the two product categories—the carrier division, which sold optical and other gear to phone companies, and the enterprise division, which sold gear for internal use within corporations or in companies' private networks. "The carrier group always had too much power. The key decisions about funding, acquisitions, and direction were always controlled by people who didn't really understand the market dynamics or direction," said Roberts. The carrier group's power was obvious from its performance in 1999, when 76 percent of Nortel's $22 billion in sales came from the carrier division, while the enterprise group could drum up only $5.4 billion in sales.[3]

The creation of two groups also created a schism between the haves and the have-nots. The enterprise group sales personnel weren't making as much in commissions as the ones who were hawking optical gear. The reorganization prompted some chaos within the company, as employees rushed to join the carrier side of the business, which seemed to have more opportunities for career and salary growth. "It appeared a lot easier to ride the CLEC (competitive local exchange carrier) boom than to compete head-to-head with Cisco in enterprise accounts," said Roberts, who quit and started his own company. "I think suddenly you had this company that was seeing its revenues double every year, and they were a bit starry-eyed about the whole thing," said another executive who sold his start-up to Nortel in 1999. In the end Bay slowly withered on the vine. At the end of 1999, Nortel boasted in its annual report, "Nortel is and must remain a growth company. Our marketing and sales teams are organized to focus on customer segments with high growth potential."

The company's new growth focus came courtesy of Wall Street. For the longest time, Nortel had been ignored by Wall Street analysts, who routinely overlooked the company in favor of Lucent or Cisco. Whenever they saw John Roth, they saw an engineer with a bad haircut who was nothing like the gregarious Lucent CEO Rich McGinn or the poster boy for high tech, Cisco CEO John Chambers. It wasn't until 1999 that Nortel caught the attention of Wall Street shills. Apparently, analysts encouraged Nortel to highlight its optical achievements so it could move the stock higher. Krish Prabhu, former chief executive of equipment maker Alcatel America, recalled that Wall Street analysts, including some at top investment banks, encouraged him to change his tune to optical and get higher valuation for the stock. Unlike Prabhu, Roth listened to their advice because the company needed a high-flying stock.

Nortel needed to increase its stock price for two reasons—it wanted to make acquisitions, and it wanted its managers to get rich so it could retain them. Nortel executives were being vigorously headhunted during what was the biggest bull market for technology stocks, because they were professional, technically savvy, and had solid, old-school executive virtues. Hundreds of little networking equipment companies that were cropping up in Silicon Valley needed to staff up. Nortel engineers, thanks to their breakthrough work in optics, were being lured with the kind of offers that are normally reserved for baseball's free agents. Nortel fought back with stock options.

Sure, Nortel's stock options were attractive, but the culture it created was one in which personal gains overrode the desire to do the right thing.

A September 2001 report from Joseph Fuller and Michael Jensen of the Monitor Group, a Cambridge, Massachusetts, consultancy, wrote of the dangers that came with overvalued companies, like Nortel, that were "reluctant to admit revenue shortfalls" and had "an unwillingness to give up the overvalued equity currency that gave managers leeway to make unwise, value-destroying investments." In sum, "managerial unwillingness to bear the pain of correcting the market earlier led to even greater pain down the road," Fuller and Jensen wrote. "CEOs are in a difficult bind with Wall Street. They can stretch to try to meet Wall Street's expectations, or prepare to be punished if they fail. An overvalued stock sets in motion a variety of organizational behaviors that often end up damaging the firm."

Nevertheless, kowtowing to the analysts worked—Nortel's stock, which was traded at around $9 in early 1997, traded in the $20 range in mid-1999 and reached $50.32 on the last day of that year. Nortel was trading at $82.51 in September 2000, six months after technology stocks began to struggle.

In December 2000, Roth was *Time Canada*'s "Newsmaker of the Year" and one of *Forbes.com*'s top 10 technology executives. "Since capping a steady rise through the ranks in engineering and operations to become chief executive in the autumn of 1997, Roth has overseen a corporate facelift that would make Zsa Zsa Gabor proud. He has engineered some 16 acquisitions while putting the pedal to the metal internally to transform Nortel from a simple telecom equipment provider into a global brand name identified with the Internet," gushed *Forbes.com*.

Nortel 1, Lucent 0

Since 1997, a lot had happened at Nortel's more esteemed rival Lucent. On October 6, 1997, Richard McGinn had been made chief executive of the company, a job he'd wanted for years. At age 51, McGinn had been handed the keys to a kingdom that included Bell Labs, the crucible of invention and innovation that had given the world the transistor and the phone switch.

McGinn had been Lucent's president and chief operating officer, and earlier had a good shot at being the CEO, but AT&T senior management brought in Henry Schacht, who had just retired as CEO of Cummins Engine Company, to run Lucent when it was spun out of AT&T.

This left McGinn fuming. When McGinn did become Lucent's CEO, it was the pinnacle of his career, which he had begun as an account executive in 1969 at Illinois Bell. With $27 billion in sales, and about $1 billion in profits, Murray Hill, New Jersey–based Lucent was one of the biggest technology companies in the world. More than a century old, Lucent's logo, a crudely drawn zero, said: If the abacus starts with a zero, telecom starts with Lucent. And now, finally, this 100-year-old force was McGinn's to control.

At the time of his appointment, Lucent had been on a roll. Unshackled from AT&T, Lucent had gone public on April 4, 1996, barely two months after the Telecommunications Act of 1996 had been signed by President Clinton. Lucent raised $2.619 billion, making it at the time the largest initial public offering in history. Lucent's shares were the most sought after on Wall Street at the time of its offering, and McGinn and other Lucent executives had substantial windfalls.

The timing was impeccable. Broadband hype was going into overdrive, and Internet was the mantra on Wall Street. Lucent's scientists had developed devices that used a combination of digital and optical technologies and could send data from New York to London at light speed. Lucent had developed a system that had one-fourth the oomph of a Nortel device that would come out later, but in 1996, Lucent's device was the best thing on the market and did the job. In a few short months, Lucent went from being a sleepy old equipment maker to a name worth watching. It had a virtual monopoly on the business.

Thanks to its AT&T genes, Lucent's stock was one of the most widely held in America. At the time of McGinn's appointment, Lucent traded for (split-adjusted) $16.39 a share. Before its IPO and separation from AT&T, Lucent's sales had primarily come from its parent and regional Bell operating companies. The company struggled to gain traction with companies like MCI and Sprint, who competed with AT&T in the long distance business and were reluctant to throw money to their stolid competitor. These companies opted for gear from Nortel and other established players such as Alcatel of France and Siemens of Germany. But the spin-off helped Lucent overcome the AT&T stigma, and sales zoomed. Of course, it helped that there were a lot more new phone companies to sell to!

At the end of fiscal 1998, Lucent's sales were up 13 percent over the previous year. In 1999, the company posted revenue of $30.617 billion, an increase of 26 percent over the previous year. Its sales would top out at $33.8 billion in fiscal 2000—a 57 percent increase in revenue since 1997. Lucent's net income was growing even faster—an increase of 127 percent, from $470 million to $1.1 billion, between fiscal 1997 and 1998, and growth of 350 percent—to $4.8 billion—between fiscal 1998 and fiscal 1999. The company, which beat profit estimates 15 quarters in a row, was a must-have in any investor's portfolio.

Between October 1996 and October 2000, Lucent went on an acquisition binge, buying 38 networking companies for about $43 billion. It wanted to be Cisco and Nortel—it just wanted to take over the world. The biggest one was Ascend Communications, an Alameda, California–based company that Lucent bought for $20 billion in January 1999. All these deals then looked cheap. "At one point we had 110,000 employees—I did not work for a company; I worked for a country," said Rita-Eileen Glynn Smith, a former marketing executive.

But numbers never told the real story of Lucent. It was a company living in the past, plagued by incompetence, complacency, and arrogance. Lucent made the classic mistake that many successful companies have made—it became complacent in its success and did nothing to develop the next generation of optical technology, thus falling behind Nortel. That became a costly mistake for a company that was worth a cool $215.18 billion at the height of the broadband mania.

McGinn's Folly

Back in October 1996, Lucent decided that it would stick to a technology that could send data over fiber-optic networks at the speed of 2.5 gigabits per second. Also known as OC-48, the technology was fast enough to transmit the contents of an entire DVD motion picture in two seconds. Nortel, which had lived in Lucent's shadow for years, had already announced its intent to sell an OC-192 system, which was four times faster than what Lucent had to offer.

Lucent engineers lobbied hard for an OC-192 product, but their pitch was nixed in the executive suite. Executives dismissed Nortel's new technology as an untried fad and later paid the price for it. It was ironic that McGinn, a sales and marketing guy, was making fun of Roth, an engineer. In the middle of 1999, Nortel had 43 percent of the optical long-haul equipment market, while Lucent had a mere 15 percent, according to the Dell'Oro Group, a market research firm. (Optical long-haul equipment carries your video clips, digital photos, and e-mails between cities and countries.)

Meanwhile, Rich McGinn, even as late as 1999, was busy playing the visionary—granting press interviews, speaking at conferences, and pretty much doing everything except minding the house. In 1998, when Lucent's market capitalization exceeded that of Ma Bell, McGinn was part of the "digital elite." He described Lucent as an Internet start-up. Yeah, sure, and Zima is an adult beverage! McGinn wanted to make the 103-year-old company behave like a start-up, but that was an impossible dream. The company was full of people who were accustomed to the slow-moving life at AT&T. They dismissed competition from companies like Alcatel, Motorola, and Nortel with a collective flick of their wrists.

Even in the dog days of 2001, the company never got over the fact that it used to be the center of innovation. "Bell Labs is the engine for this entire company. Everyone else has been content to rely on M&A activity to acquire technology, but no one has the kind of research organization we have," Bill O'Shea, executive vice president of corporate strategy, would later say.[4] Did he forget that Lucent had spent almost $20 billion on 38 acquisitions since it went public? Or did O'Shea forget that his master McGinn, in an interview with *Red Herring* in August 1998, had boasted that its "biggest advantage kicks in at midnight EST on September 30 [1998]" when "the company can finally use its massive market capitalization to make significant acquisitions." (Accounting rules prevented Lucent from doing deals for stock, using the "pooling of interest" method for three years after the reorganization. For Lucent the deadline was September 30, 1998.)

On McGinn's watch, the company went from 120,000 employees to 160,000, made 38 acquisitions, and collapsed. The biggest problem was the reorganization instituted by him in 1997—he divided Lucent into 11 business groups, and these groups often ended up competing with one another. (Others blame the management for not understanding the evolution of telecoms and not having modern technologies.) Already slow in getting out the blocks with its version of OC-192 optical technology, Lucent was mired in a bureaucratic mess and was suffering from serious political infighting. Also, while a competitor like Cisco could pitch, sell, and install a product in two days, Lucent took almost a week to make a presentation to the customer. Customers soon got disgusted and fled to competitors, which resulted in the erosion of Lucent's market share. (See Table 10.2.)

Disgruntled employees made a beeline for the nearest exit! Instead of trying to stem the flow of executives, Lucent resorted to suing former employees. In mid-1999, Cisco Systems, the hard charging rival, opened a new manufacturing plant in Salem, New Hampshire, about a half hour from Lucent's facility in North Andover, Massachusetts. Ten Lucent employees quit and joined Cisco. "Instead of simply suing Cisco, Lucent sued all ten of the former employees individually for alleged breach of contract (even

Table 10.2   Fat and Foolish

Lucent bought too many companies for too much.

Announced Date	Target	Size ($mm)
September 1997	Octel Communications Corporation	$ 980.0
October 1997	Livingston Enterprises, Inc.	650.0
January 1998	Prominet Corporation	200.0
April 1998	Optimay GmbH	65.0
August 1998	Lannet Data Communication	117.0
October 1998	Quadritek Systems, Inc.	55.0
November 1998	Yurie Systems, Inc.	1056.0
March 1999	Enable Semiconductors— Ethernet LAN Division	50.0
June 1999	Ascend Communications	24000.0
June 1999	InterNetworking Systems	24.0
July 1999	Nexabit Networks	900.0
July 1999	Mosaix, Inc.	145.0
October 1999	International Network Services	3700.0
November 1999	Xedia Corporation	246.0
November 1999	Excel Switching Corporation	1700.0
February 2000	SpecTran Corporation	99.0
April 2000	Ortel Corporation	2798.0
April 2000	Agere Systems, Inc.	443.0
May 2000	Chromatis Networks, Inc.	4505.0
June 2000	Herrmann Technology, Inc	438.0
July 2000	Spring Tide Networks, Inc.	1346.0
Total	**21 targets**	**$43,517**
	$2072.2mm average price paid	

Source: Capital IQ

though, according to Lucent's own employee guidelines, it promises not to sue its employees for going to work with a rival)," wrote *Red Herring*.[5]

Then, in the summer of 1999, sales began to stall as customers got more choices from aggressive competitors such as Nortel, Cisco, and smaller players like Ciena. This had a debilitating effect on Lucent's troops. Executives were leaving at an alarming pace. The senior management never tried to stem the flow. Instead they saw it as a sign of rejuvenation. "Thirty percent of the senior executives who were at Lucent at the time of our IPO are no longer around," said McGinn in an interview in June 1999.[6] That alone should have been a red flag—why were executives of such a fast-growing company quitting in the middle of a boom market? Perhaps they knew about the problems that lay ahead.

In a September 1999 meeting in Nuremberg, Germany, Lucent management pushed its sales team to sell products now and "fix problems later."[7] Worried that the stock would fall on news of the slowdown, Lucent started banking heavily on a financial technique called vendor financing to shore up sales and boost revenue.

My Vendor, My Banker

Vendor financing had been used by telephone equipment makers for decades as a method for breaking into new accounts or introducing new technology. This is how it works: A little phone company goes to Lucent and begs for their help. Lucent figures these guys have a decent business plan, and decides to extend them a loan of $100 million to buy all the gear they need to get started. Loaning money in this way is hardly a philanthropic gesture, because the quid pro quo is that the little start-up has to buy all its gear from Lucent and, over a period of time, pay back that loan. Back when Sprint was finding its footing as a long distance carrier, Nortel was one of the companies that provided vendor financing for the company, helping Sprint become a major telecommunications player. The key difference between financing in the early 1990s and the late 1990s, as a former Nortel finance executive explained, was that when Nortel first began using the financing scheme, the due diligence would take months, and companies had to meet very strict criteria to get vendor financing from Nortel. However, in the late 1990s, vendor financing became a sales tool. Not only would Lucent, Nortel, and Cisco provide vendor financing for equipment, but they would also invest in some of these companies as well.

In the beginning, the amount that Nortel was financing its customers paled in comparison with Lucent, which was openly financing companies, many to the tune of $2 billion. It was all due to Lucent's desperate bid to grow sales and meet Wall Street's unrealistic expectations. Trying to boost $30 billion in sales 20 percent every year is a near-impossible task. But the future of Lucent and its options-rich executives was predicated on the company's stock, which would only rise if sales numbers grew.

According to independent estimates, company filings, and press releases, Lucent, Nortel, and Cisco had committed roughly $8 billion, $4.5 billion, and $2.5 billion, respectively, in vendor-financing telecom start-ups. At the end of 2000, McKinsey & Company estimated that the nine global telecom giants—Lucent, Nortel, Cisco, Ericsson, Alcatel, Motorola, Nokia, Qualcomm, and Siemens—were carrying loans of $25.6 billion, about 123 percent of their pre-tax earnings in 1999. In other words, a large portion of the sales these companies were boasting about were actually loans.

Analyses of numbers collected from various sources show that Lucent and, to a lesser extent, Nortel were simply buying sales through vendor financing. In 1997, Lucent had vendor-financed companies to the tune of $1.9 billion (versus 1997 sales of $21.5 billion), and in 1998 that number rose to $2.6 billion (versus 1998 sales of $24.4 billion). However, by 1999, Lucent's vendor financing loans stood at $7.2 billion versus its sales of $30.6 billion in 1999. Nortel's numbers were smaller but no less significant. Vendor credit issued by the Canadian giant was $1.6 billion in 1999, three times as much as the $573 million issued in 1998.

Lucent and Nortel were giving loans to companies even a neighborhood loan shark would have turned down. Take, for instance, Lucent's loan to Fidelity Holdings. No, not the mutual fund giant, but a small Queens, New York–based company that owned car dealerships and a sludge treatment business! It also owned a venture called IG2 that sold Internet, phone, and video service over existing phone lines. Fidelity Holdings had revenues of about $1.3 million in 1999, but that didn't deter Lucent from giving IG2 financing of $400 million for equipment.[8] Don't bother looking for IG2.com or the web site for its parent company today. The domains are for sale, and IG2 is out of business.

At Lucent, Carly Fiorina, then the head of global carrier sales, was a champion of aggressive growth and vendor financing. Fiorina, who is now Hewlett-Packard's CEO, was responsible for a large portion of Lucent's sales and was one of the key people who negotiated a $2 billion financing deal with Winstar, one of the primary beneficiaries of Lucent's largesse, back in 1997. "She [Fiorina] intimated to employees and outsiders that Wall Street would generously reward companies that emphasized and delivered robust revenue growth," wrote *Fortune* magazine.[9] Even after Fiorina left for Hewlett-Packard in July 1999, the vendor financing practice continued. "Lucent is financing in a crazy way," Rick Gilbert, CEO of Copper Mountain, a Lucent competitor, told the *Industry Standard* in early 2001.[10] "If a company asks for financing, all we can do is send them to GE Capital or someone like that. Lucent can do it themselves, and on terms no one else would dream of giving away."[11]

Lucent's hardball tactics forced other competitors, such as Nortel and French equipment maker Alcatel, to respond with their own vendor financing programs—sometimes with disastrous results. Alcatel, in a desperate bid to shore up its market share, agreed to provide $700 million in financing to Greg Maffei's 360networks in November 2000. 360 went bankrupt later, leaving Alcatel on the hook.

Nortel, which had always suffered from an inferiority complex, was most aggressive in its pursuit of business and fought tough with Lucent. In 1998, Net2000, a phone company with delusions of competing with the likes of Verizon, had failed to raise money through a junk bond offering and was running out of options when both Lucent and Nortel showed up with bags of money. Lucent and Nortel fought tooth and nail to win the account. To win, all the Canadians had to do was offer a $180 million deal and buy $30 million of Net2000's convertible preferred stock.

"Just like everybody else, we got caught up in the heat of the moment," confessed a former Nortel executive who was closely involved with the company's vendor financing efforts. With Lucent, Alcatel, Cisco, and pretty much everyone ready to offer vendor finance, Nortel was running out of options, he said. For many years Nortel had been methodical in its due diligence when it came to vendor financing and always took a bet when executives believed they were investing in a company with good management. But in 1999, Nortel threw caution to the winds.

By late 2000, Nortel decided to provide vendor financing for a FreeDSL service started by the Steelberg brothers, whose previous claim to fame was selling AdForce, the Internet advertising service, to CMGI, an Internet holding company in Massachusetts, for $500 million. The FreeDSL service from Winfire basically offered a few high-speed broadband connections for people who did not mind watching ads on their personal computers. The former finance executive recalled, "At the time, if you said anything telecom, you could get funding."

Why was it so easy? Because Wall Street was willing to lend money to guys like Nortel and Lucent at ridiculously low rates, who could then take this money and fund little telecom start-ups and boost their own revenues. Think of the little telecom start-up as a fresh graduate with a $30,000 a year salary, looking to rent an apartment in New York, which, unless Daddy co-signs, would be impossible to get. It was the same case with the telecom start-ups. "There were more and more entrants in this business, and that was scary, but we had to use vendor financing to win business. There was no way you could get off the treadmill. It was like you had the tiger by the tail," said the former insider.

By 2000, as other avenues of capital started drying up, vendor financing became the only way for struggling carriers to raise new money, which explains why vendor financing dollars ballooned in 2000. Lucent, Cisco, and Nortel came to start-ups' rescue, but it was a bad move—Lucent's total bad loans increased from 2.6 percent of total loans in 2000 to 60 percent of total loans in 2001, and Nortel's bad loans rocketed from 25.5 percent in 2000 to 80 percent in 2001. "For the whole industry, that was table stakes. In fact, we were criticized in some measure for not being aggressive enough, that guys like Lucent and Cisco were way more aggressive than Nortel, and Nortel was being criticized in some cases for not being aggressive enough," Roth told the *Toronto Star*, defending his decision to be an aggressive vendor financier.[12]

The Big Zero

Toward the end of 1999, with its stock slipping fast at $57.14 a share, down from an all-time high of $62.69 a share (on December 20, 1999), Lucent's management, including McGinn, were privately worried. There was little chance of new millennium festivities for these guys. Over the past few weeks, the company had been offering steep discounts to its big customers in order make its revenue targets for the quarter. It was a short-term strategy with absolutely no regard for the future. McGinn obviously had never heard the story of the goose that laid the golden egg.

Apparently, in December 1999, McGinn told delegates from the Communications Workers of America that "since the IPO, four years earlier, the company borrowed money to meet each payroll with the exception of one quarter."[13] Morton Bahr, the group's president, retold this story in a speech and added, "This was astonishing, as the shareholders were not aware of the precarious position of the company as their shares continued to escalate in price. This was shady bookkeeping at its best. It was later learned that Lucent was recording as revenue received orders for material not yet delivered and work not performed."[14]

Somehow, word of this got to the stock market, and the stock skidded to $39.62 in the first week of the new year. Over the next nine months, the company would issue three profit warnings and would keep trimming its estimates. In the third quarter of 2000, Lucent finally fessed up: Its optical networking business was down 5 percent from the same quarter a year earlier. It was a problem not unique to Lucent, as we learned later. In October 2000, Nortel's Roth delivered the same bad news to Wall Street—the sales of its optical networking equipment had fallen short of forecasts. He dismissed it as a one-time event, blaming a shortage of people who could install this equipment at carriers.

While Roth was still bullish about the prospects for his company, judgment day for McGinn was just weeks away. Lucent employees were completely demoralized—they were working six-hour days, taking three-hour lunches, and were looking for jobs elsewhere. The stock had sunk to the mid-$20s by September 2000, and their stock options had basically lost all value; and the slide wasn't over.

On October 23, 2000—almost three years and two weeks to the day since he became Lucent's chief executive—Rich McGinn was fired, though not before the company set him up for life with a severance package worth $12.5 million. This included $5.5 million in cash in addition to the $20 million or so he got in salary and bonuses between 1998 and 2000.[15] He also got some 17 million shares of Lucent! Incompetence paid off handsomely for him. His total take for reducing Lucent to a nonplayer: about $38 million, and that is without the value of the shares.[16] Today, McGinn is a partner with RRE Ventures, a New York–based venture investment group that invests in technology and telecom companies.

A few weeks after McGinn was fired, details of all sorts of shenanigans started coming to light. The company discovered that some salespeople were "cooking the books." Lucent had to restate its earnings for the quarter ending September 30, 2000, by $259 million and revenues by $679 million. In a conference call with Wall Street analysts, interim chief executive Henry Schacht, who was McGinn's predecessor and who had stepped in again after McGinn was fired, said that a Lucent sales team had drawn up a phony sales document that resulted in a bogus sale of $74 million. In another instance, the company recorded $28 million in sales for a product that was not even ready.

In a meeting with Lucent executives at a Hilton hotel in Parsippany, New Jersey, Schacht chastised the company for "stretching revenue-recognition practices" and "putting off write-offs," according to an internal memo, which went on to add that there was "a breakdown of the basic processes of the company. We tried to run faster than proved possible, and we tried to run the company too hot."[17] (See Table 10.3.)

In the two years since McGinn left, Lucent has become an incredibly shrinking company—it has spun off many of its divisions, like Agere Systems and Avaya, and laid off over 60,000 employees. For fiscal 2002, it reported sales of *$12.3* billion (a drop of 42 percent over 2001 revenue) and losses of *$11.8* billion. It is now contemplating a reverse stock split to stay above *$1* a share so it doesn't lose its position on the New York Stock Exchange. The U.S. Securities and Exchange Commission is looking into the company's accounting practices.

It's easy to see why things went wrong at Lucent. In an interview with *Fortune* magazine, Henry Schacht remarked, "Stock price is a by-product; stock price isn't a driver. And every time I've seen any of us lose sight of that, it has always been a painful experience."[18]

Masters of Their Optical Domain

Nortel's optimism lasted a little longer. Even though it was increasingly being saddled with vendor-financing loans, and even as the demand for optical networking equipment was declining, it was still being touted as a sexy stock play. "Everybody loved working there. The place was buzzing. It was an atmosphere of success," a Nortel employee told the *Toronto Star*.[19]

Nortel is based in an industrial park in Brampton, Ontario, about 45

the Fast Lane
pushed it to the limit and then over the edge.

	12 months Ending /30/1996	12 months Ending 9/30/1997	12 months Ending 9/30/1998	12 months Ending 9/30/1999	12 months Ending 9/30/2000	12 months Ending 9/30/2001	Las 12 mc Endi 9/30/
			In Millions of U.S. Dollars (except per-share Items)				
	5,859.0	27,611.0	24,367.0	26,993.0	28,904.0	21,294.0	12,321
	(25.9%)	74.1%	(11.7%)	10.8%	7.1%	(26.3%)	(4.
	224.0	449.0	1,065.0	4,789.0	1,219.0	(16,198.0)	(11,75.
	1.4%	1.6%	4.4%	17.7%	4.2%	(76.1%)	(9.

rket Guide

minutes northwest of Toronto. Even locals have a tough time finding the sprawling campus, which was dubbed "The City" by 3,500 employees who worked there, and which is overshadowed by a Ford manufacturing plant.[20] There was a huge food court, landscaped grounds, jogging trails, and all the perks normally enjoyed by employees of Silicon Valley companies like Cisco Systems. Roth and others at Nortel clearly suffered from Cisco-envy. (It was a malaise not unique to Nortel; even Lucent wanted to be Cisco.) "Cisco has a bigger market cap, Lucent has more in sales, Ericsson is ahead of us in wireless. If we rise above them, we will invent a [market] leader to fight against," Roth told *Forbes*.[21]

Nortel was completely focused to out-Cisco the Silicon Valley giant. And why not? Cisco executives at the time were a legend in the business—hardball sales tactics and millions of dollars in options had made them among Silicon Valley's wealthiest. In the late 1990s, there were more BMWs in the Cisco parking lot than anywhere else in Silicon Valley, including the local BMW dealership, and it wasn't unusual for the company to have sales conferences in exotic places. Nortel, it seemed, was ready to do anything to match Cisco's culture. By 2000, Nortel was spending lavishly on company events and conferences. Monaco played host to the company's 2000 sales conference. It was one of many extravagant events held that year.

Former executives say that the free-spending ways came to Nortel courtesy of Clarence Chandran, the chief operating officer. The son of an Indian army officer and a self-described "army brat," Chandran immigrated to Toronto, Ontario, with his family in 1969, when he was 15. For the young Indian it was an amazing experience. "There was this incredible excitement and I was fortunate to some degree that, being in the army, we used to travel and live in strange places," he told *Rediff.com*, an Indian Web magazine.[22] "I was comfortable in a foreign environment but it was still tough to find my way around. [Canada] was a very young country and the charismatic government led by Prime Minister Trudeau was trying hard to promote bilingualism and ethnicity."

At age 20, Chandran started working for Shell and joined Bell Canada five years later. In 1985, he joined Nortel, and for the next several years slaved away in relative anonymity as a middle manager in the technology backwaters of Latin America and Asia. But his hard work paid off, and in February 1990, he was given Nortel Networks' worldwide Award of Excellence. It proved to be a turning point in his life.

In the early 1990s, when Nortel's sales were rising, Chandran had hitched his wagon to John Roth. Insiders say that when Roth was named chief executive, Chandran's rise to the top was almost assured. Chandran's projects included a foray into fixed wireless equipment and new products Nortel had promised to develop for Alex Mandl's Teligent, the failed carrier. In June 2000, Chandran was made president of Nortel's Global Carrier and Service Provider business, and finally, on June 27, 2000, he was named the chief operating officer and the heir apparent to the CEO throne.

Having made the right bets on new markets like wireless and optical, Roth was handing over day-to-day control to Chandran, who, unfortunately, lacked Roth's eye for detail and his engineering background. He was more a bureaucrat, not ready for the job that was thrust on him. "Roth had made Nortel into a sexy company," recalled an executive who worked in Nortel's Silicon Valley office at the time. Roth wanted to get out at the very top. "But Clarence could not make big decisions, and he was so caught up with growth that he forgot the fundamentals."

One of Chandran's first acts as COO was to shift responsibility for profit and loss statements from specific business units to the sales teams. As a result, salespeople were put in charge of the company's profitability and were given control of budgets. Talk about committing financial hara-kiri! "There was no accountability, no controls, and no one had the right numbers and financial picture at the company," said one insider. At its peak, Nortel had 500 vice presidents who made six-figure salaries, flew business class, and lived a cushy life. They could even use the company planes freely!

In most industries, salespeople have low salaries and make the bulk of their income from commissions. The need to focus on their own commissions didn't instill fiduciary responsibility in Nortel's salespeople. Many lacked fiscal discipline to prevent excessive spending, aggressive use of financing, or unjustified price discounting—all problems that would begin to plague Nortel. "Sales teams were busy getting their commissions, and they were told, get the deal and grow the revenues," said one former Nortel marketing executive. "Decisions were made based on hype, greed, and ego." Even though in mid-2001 Nortel's business units regained control of profit and loss statements, it was a case of too little, too late.

Nortel's senior management was still living in a fantasyland. "Everyone is coming after us. We have to take winning the war on talent to the next level," said Roth, and bragged that Nortel was paying bonuses, sometimes equal to an annual salary, twice a year.[23] It was clear that Roth, in his desire to grow revenues and keep Nortel ahead of the pack, was losing all perspective. Money just became a tool to push his sales teams closer to the precipice.

Using Nortel's red-hot stock as currency, Nortel imitated Cisco's acquisition strategy and went after hot start-ups, paying billions of dollars for them. Nortel spent $30.6 billion on 18 companies between 1997 and 1998, hoping that, like Cisco, acquisitions would help it grow revenues and move into new markets. Cisco's merger and acquisition strategy was quite simple—it bought companies with great technology that could be quickly mass produced and sold by its crackerjack sales force. With few exceptions, Cisco exercised a lot of self-restraint when it came to the final price tag.

In early 2000, Nortel considered buying ArrowPoint Communications, a maker of web switches, but decided that ArrowPoint's asking price of $1 billion was too much. Shortly thereafter, ArrowPoint went public, and Cisco bought the company out for $5 billion. "Of course, this set everybody at Nortel off, thinking that Cisco knew something we didn't know. The reality was, Cisco was overpaying, too. The stupidity was so high," recalled Dave Roberts. Partly in reaction to that experience, Nortel spent $8 billion to acquire Alteon WebSystems, an ArrowPoint rival, in July 2000. The deal was considered overvalued, and Nortel never recouped the cost of that deal. "At its height, I don't think Alteon ever broke $50 million in sales," said Roberts. Today, even by the most generous estimates, Alteon is only generating $25 million in quarterly sales.

Other insiders agreed and said that Nortel was forgetting two things—it had a bloated infrastructure and high fixed costs. It needed to pay 100,000 employees, while Cisco had half as many people. Even at its peak, Nortel never could match Cisco's sales of $500,000 per employee. And unlike Cisco, Nortel was never able to milk its acquisitions as well, or sometimes the company simply bought second-tier start-ups.

But as Nortel continued to pamper its sales team, many of whom were pencil pushers in disguise, quite a few of their customers were getting some sort of funding from Nortel. In the fall of 2000, Nortel organized a conference for the North American sales team. As part of this event, about 1,000 salespeople were flown in to Vancouver, the Canadian Olympic sailing team was booked, and regatta races were held in Vancouver Bay, along with extravagant dinners and receptions during the three-day event.

On another occasion, the 150-strong Optical and Access Sales team were hosted in Chicago in late 2000. The main event of the two-day conference—a black tie catered reception and dinner, followed by a drawing for a Jaguar convertible—was held at the Field Museum, which overlooks Lake Michigan. Guests entering Stanley Field Hall were greeted by gleaming white marble interiors, skylit vaulted ceilings, stately columns, a magnificent grand staircase fit for a princess, and "Sue," the largest and most complete Tyrannosaurus rex fossil yet discovered. Renting even parts of the museum could set you back by as much as $8,000 a night. Nortel rented the entire museum, which could easily hold more than 5,000 people.

The forces of reality would catch up with Nortel a few months after McGinn was fired at Lucent, but up until then it was a wonderful world for Roth and his sidekick, Chandran. Many of Nortel's customers were having trouble generating business—which meant their equipment purchases, many of which were already financed by Nortel, were slowing. But still Nortel shares were in a gravity-defying mode. Roth and Nortel were on the cover of every major business magazine, and Roth's every utterance was viewed as the gospel of optical gods. Fawning press reports made Nortel seem almost invincible. It was the perfect time for Roth to cash out. On August 9, 2000, he exercised 360,000 options at $7.74 a share and another 90,000 options at $11.93 a share. He sold the total 450,000 shares the same day for a total profit of about $40 million. In 2000 alone, Roth had pocketed a cool $100 million—including $1.1 million in salary plus a bonus of $5.6 million and an additional $5.6 million from another incentive plan.[24] Making money and playing god—Roth was on top of the world. (See Table 10.4.)

Wall Street analysts ran out of superlatives. But Paul Sagawa, a former AT&T executive and now an analyst with Sanford C. Bernstein, wasn't buying it. He had spent two months in the summer of 2000 interviewing 60 carriers in the United States and Europe and had come to the conclusion that there was a slowdown coming, and that it would affect Nortel

4 Easy Come and Easy Go

enues grew fast; they shrank faster when the optical winds turned.

	12 months Ending 12/31/1997	12 months Ending 12/31/1998	12 months Ending 12/31/1999	12 months Ending 12/31/2000	12 months Ending 12/31/2001	Last 12 months Ending 9/30/2002	12 months Ending 12/1/200. (estimated)
	In Millions of U.S. Dollars (except per-share Items)						
	15,449.0	**17,575.0**	**19,628.0**	**27,948.0**	**17,511.0**	**11,496.0**	**10,381.4**
	20.3%	13.8%	11.7%	42.4%	(37.3%)	(48.3%)	(40.7
	712.0	**(1,250.0)**	**(351.0)**	**(3,470.0)**	**(27,302.0)**	**(5,163.0)**	
	4.6%	(7.1%)	(1.8%)	(12.4%)	(155.9%)	(44.9%)	

ɔital IQ, Market Guide

quite badly. He was ridiculed by other analysts, mutual fund managers, and by individual investors, who vented their anger towards Sagawa on online message boards such as Raging Bull and Yahoo! Finance. Pity Sagawa couldn't counter them on message boards!

When Sagawa's report crossed my desk, I called him. It made perfect sense, for at the time my sources were telling me that bandwidth prices were in free-fall, and many carriers were having trouble closing sales even at highly discounted prices. Comparing notes with Sagawa, it became obvious that almost everyone was ignoring the coming crisis. "Telecom, like all businesses that require massive investments, is cyclical. We have had a long up cycle, and I think we are going into a down cycle," Sagawa told me in an interview for a story published in *Red Herring*.[25]

It was clear that for the boom to continue, there needed to be a more rational bandwidth pricing structure, but that wasn't feasible, given the amount of excess capacity that existed in the fiber-optic network business. After Sagawa's research report was published, Nortel publicly derided him and put its entire public relations team on the job to contain the damage. Nortel's then high-powered PR agency Fleishman-Hillard sent e-mails to technology reporters, dismissing Sagawa's analysis as "hand-wringing by worried analysts."

Like Sagawa, Susan Kalla, at the time an analyst with the small New York–based brokerage firm Bluestone Capital, had come to similar conclusions. Other telecom industry watchers had a different viewpoint. "What will the carriers do, not spend on equipment and try and meet the demand for bandwidth, which is exploding?" asked Paul Silverstein, networking equipment analyst with Robertson Stephens, an investment bank that has since gone out of business. It did make you wonder—how did these financial mavens forget the very basic principles of economics? If people don't buy cars, demand for spare parts and then for gas vanishes. The fiber-optic business was no different, but everyone turned a blind eye.

Carriers were cutting orders on new gear, mostly fiber-optic gear. That worried Anil Khatod, who at the time was the president of Nortel's global Internet solutions. In October 2000 he made a presentation to the executive team suggesting they should tone down the forecasts.[26] But no one paid attention. On October 24, 2000, after the markets closed, Nortel announced its earnings: The company took in about $7.31 billion revenue, $200 million shy of the target Wall Street had set for the company. The Canadian giant was immediately punished for missing its revenue target of $7.5 billion. In after-hours trading, shares of Nortel sank like the *Titanic*—down almost $12.56 to $50.75, after it had already lost $3.63 a share in trading on the New York Stock Exchange ahead of the October 24, 2000, earnings announcement. The next day was even worse—Nortel stock fell almost 30 percent, closing at $45 a share. Within 24 hours, almost $55 billion in market capitalization was wiped out. Fiber optics stocks such as JDS Uniphase, Sycamore Networks, and Corvis, which had defied gravity for much of 2000, also fell to the ground. In its own strange way, the age-old axiom, "the stock market knows the future first," was coming true.

Roth, however, thought otherwise and lashed out at the very same investors who had made him the king of telecom. In an interview with *Chief Executive* magazine,[27] he lamented, "I'm an engineer and it reminds me of a type of feedback system called the bang-bang control theory—it's either all on or all off. That's what the market does; it's either pedal to the metal or put on the brakes, but it's nothing in between. The market went into a tailspin because they wanted us to grow 52 percent. But we never said we would grow 52 percent. I think people watched our 150 percent growth rate in the first half and assumed that was the growth rate of the market."[28] But it was a bad excuse, since Nortel's management were the ones providing guidance on the company's growth to Wall Street analysts and telling them what kind of growth to expect.

Despite analysts' growing concern and Nortel's own lowered forecasts in late 2000, Roth continued to promise growth for 2001, both in its revenues and earnings per share. Twenty percent revenue growth was not going to be a problem. On December 14, 2000, he told the analysts, "We continue to expect to grow significantly faster than the market, with anticipated growth in revenues and earnings per share from operations in the 30 to 35 percent range."[29]

Two months later, Roth would come to regret these words. On February 15, 2001, while on a trip to Europe, Roth got a phone call from his office back in Canada. The gist of that call: We are in trouble. Executives from Nortel's head office told him that the sales were tanking. "It was like hearing you lost a relative. It was like 'Holy shit'—I think those were the words," he later told the *Toronto Star*. Nortel later issued a press release and readjusted its revenue and earnings targets. Instead of earning money, Nortel said it expected to lose some—sort of like saying "Oops we made a mistake, and instead of making money, we are now going to lose a lot—almost 4 cents per share." Of course by then, Roth had had enough time to cash out of about $90 million in stock options, including the $40 million he'd already cleared—that's in addition to the nearly $20 million in salary and bonuses he had received from the time he became chief executive.

On February 16, the day after Nortel announced it would lose money in that quarter, Nortel stock plunged nearly 33 percent. The company's stock opened at $30, down $16.15 a share on the Toronto Stock Exchange. Nortel finished at $31, off $15.15 on the day on a volume of 23.9 million shares. The sell-off was so severe that the TSE had to halt trading in the stock several times during the day. It was a Black Friday for Nortel—almost $44 billion was lopped off from its market capitalization.

Soon after, shareholders filed a lawsuit alleging that Nortel executives knew about the bad news and chose not to share it with investors. The suit alleged that William Connor, president of Nortel's e-business solutions group, and Chahram Bolouri, president of Nortel's global operations, sold over $7 million of company stock after Nortel announced strong fourth-quarter earnings on January, 18, 2001.[30] Canada's major pension plans, like Canada Pension Plan and the Ontario Teacher's Pension Plan, as well as the mutual fund Altamira Equity Fund, lost a combined $1.5 billion. Much of that money came out of the pockets of widows, teachers, and firemen, who were counting on it for their retirement.

Like rats deserting a sinking ship, Roth decided to retire in May 2001, and Clarence Chandran cited personal and health reasons for his own resignation. Today, Nortel is on life support, a pale imitation of its former self.

11

THE DAN AND DESH SHOW

Evangelism comes with a price, and no one knows that better now than Gururaj Deshpande, founder and chairman of Sycamore Networks, a maker of optical network equipment. The company that at one time had a market capitalization of $51.5 billion is now struggling to stay afloat. The problems began when "Desh" the entrepreneur became "Guru" Desh, the spiritual leader of optical networking start-ups.

Up until the mid-1990s, most telecommunications networks relied heavily on semiconductor-based electronic equipment to transmit and control heavy traffic, whether in voice or data. Sycamore wanted to change that. The company's executives wanted to build high-capacity optical-networking devices that used waves of light to deliver huge amounts of voice, data, and video traffic from one end of the country to the other, through fiber-optic cable. The benefit: You could send huge amounts of data cheaply at light speed.

Desh's dream was to build a system that could provide real-time bandwidth to customers. Sycamore wanted bandwidth available as easily as flicking on the light switch. Turn on all the lamps, draw more power. But when the lights go dim, power usage goes down. For example, with such a system, major league baseball could order a 10-gigabit connection for game seven of the World Series, and then return to a half-megabit connection once the series ended. He believed customers would pay a premium for such a service. "Sycamore is not about money. It is about the challenge and building something new," Deshpande told me during the course of an interview for *Forbes.com*.[1]

With an increasing number of long-haul carriers such as Qwest and Global Crossing making plans to grow aggressively, Deshpande bet there would be huge demand for the kind of gear Sycamore would make. He believed that the new generation of broadband service providers would ultimately win. He was wrong. When the general euphoria for telecom stocks went into overdrive, Deshpande, unlike others, did not talk up the Sycamore customer base or hype the company. But Sycamore was over-confident—and picked the wrong customers.

Desh had always been an overachiever. Born on November 30, 1950, in the small south Indian town of Dharwar, in the Indian state of Karnataka, Deshpande was always a bright kid. Like most middle-class Indian families, the Deshpande household emphasized the importance of higher education. No one was surprised when Desh was admitted to the Indian Institute of Technology (IIT) in Madras to pursue a bachelor's degree in electrical engineering. The IIT network was established in the 1950s by India's first socialist Prime Minister, Jawaharlal Nehru, whose vision was that the IITs would help India produce top-notch engineers. He was right. Once students got a degree from any of the IITs, their future was assured. After graduating, Desh moved to Canada for further studies. He received his master's degree in electrical engineering from the University of New Brunswick and a Ph.D. in data communications from Queens University in Ontario.

While teaching at the university, he was lured by one of his former professors to Codex Corporation, a modem maker that was later acquired by Motorola. He was relocated to Codex's offices in Boston in 1984. "At Motorola in Toronto, I expanded a small data multiplexing division into a $100 million business. Creating that business gave me the confidence I needed to become an entrepreneur. It also gave me the freedom to work in different areas of the company to broaden my skills," Deshpande wrote in an article for *Red Herring*.[2]

■ ■ ■

Bitten by the entrepreneurial bug, Deshpande started Coral Networks in 1987. The company was developing a router to compete with Cisco, which was then only just getting started itself. A falling out with his business partner led to Deshpande leaving in 1990. The company was sold in 1992 and Desh's take from that was just $26.

The experience didn't faze him. He wanted to start another company, but had no cash to finance such a project. Although he joked about it later, he and his wife, Jayshree, another IIT graduate, were going through a rough financial patch during those years in the early 1990s. Jayshree had quit her job as a software engineer, Desh had walked out of Coral, and there was no money in the bank. The young couple decided to take a four-month sabbatical to India. "I had no money, but once you've run a company, it's very tough to work for anyone else again," said Deshpande.[3]

That's when he co-founded Cascade Communications in Westford, Massachusetts, to develop frame-relay equipment. The products would allow corporations to connect to private data networks and to share information. He invested a few thousand dollars of his own money and later got some funding from Boston-area venture capitalists. His co-founder, Wu-Fu Chen, would later become a broadband luminary.

It was at Cascade that Desh met Daniel Smith, a man who would become Desh's longtime collaborator and business partner and a close friend. A year older than Desh, Smith was born in Nyack, New York, and got his bachelor's degree in engineering from Lehigh University and an M.B.A. from Harvard. He joined the U.S. Navy after finishing school but quit to join Rolm Corporation, a manufacturer of telecom networking equipment. He then moved on to Proteon, another networking equipment maker, as the vice president of sales, before joining Cascade as chief executive. The two became a perfect foil for each other; Deshpande was the technology guru, and Smith was the one who got things done—like acquiring customers. Smith was one of those guys who worked hard at everything he did. The story goes that he worked so hard on his golf game that he once tore the muscles in his rib cage! "He sets milestones—things like: 'Get me this customer and I will get you another billion in valuation,'" Ryker Young, Sycamore's vice president of sales, told *Business Week*.[4]

Cascade quickly grew to 900 employees and $500 million in sales. It went public successfully, but sliding sales drove the high-flying stock down, and Wall Street started pressuring the young company to sell itself. In March 1997, Cascade was bought by Ascend Communications, another data networking equipment maker, for $3.7 billion. Later, Lucent

Technologies acquired Ascend for $20 billion. Deshpande and his partner Daniel Smith made about $150 million each from the Cascade sale.

■ ■ ■

Sycamore's story began in 1997, when two Massachusetts Institute of Technology researchers, Rick Barry and Eric Swanson, ran into Desh at a Christmas party hosted by Matrix Partners, a powerful Waltham, Massachusetts–based venture capital firm. Barry and Swanson worked at MIT's Lincoln Labs and had developed a mini-fiber–optic network that used mostly optical equipment. Deshpande's first interaction with the MIT duo convinced him that he had a chance to redefine the very infrastructure of the Internet and head up a company that could become a huge player in telecommunications. Desh got Smith to sign on and invested $2.5 million of his own money into the start-up, even though it had no business plan.

The yet-unnamed company was based in Tewksbury, Massachusetts. It got its name from a case of mistaken identity: One of the company's executives mistook two maple trees outside the office for sycamore trees and suggested the name Sycamore Networks. As the number of employees increased, Sycamore moved to a bigger space in Chelmsford, Massachusetts.

Desh likened the still-evolving optical networks to the early days of telephone, when telephone operators physically connected callers to one another. Until then, optical networks needed to be manually configured, and the data could run only in circles, also known in the industry as SONET rings. Sycamore would turn the optical networks into a modern phone, where you just pick up the handset and dial. It would do that by developing hardware products and combining them with intelligent software to automate the network configuration. But that required money.

Lured into the Matrix

To fund this initiative, Smith and Deshpande made a beeline for their old benefactors, Matrix Partners, which had helped them with financing for Cascade. "It took less than 15 minutes to line up the financing. Once you do a successful start-up, it becomes relatively easy to get access to venture capital," said Deshpande.[5] Matrix Partners invested $2.5 million.

Later Sycamore raised $35 million from Northbridge Venture Partners, Amerindo Investment Advisors, Bowman Capital Management, Integral Capital Partners, and Pequot Capital Management.

But it was the Matrix investment that catapulted Sycamore into the spotlight. On the East Coast, Matrix is every bit as powerful as the more well-known Kleiner Perkins Caufield & Byers on the West Coast. The dot-com mania may have started in Kleiner Perkins' offices, but the broadband boom's roots could be traced to Matrix's Waltham, Massachusetts, offices. In 1997–1999, Venture Economics, a division of Thomson Financial, estimated that Matrix was one of the most active investors in the broadband sector. At last count, in November 2002, the firm had invested in over 56 companies; Sycamore was one of its most profitable investments.

Matrix's founder, Paul Ferri, 63, is considered the god of broadband start-ups. In 1982, Ferri had started an investment partnership, Hellman-Ferri Investment Associates, and later changed the name to Matrix Partners. In the early days, the firm invested in a variety of sectors and had a mediocre record. But by the late 1980s, it had shifted its focus solely to technology companies. It was a good move, as Matrix soon had an enviable record.

Matrix's investment in Cascade Communications proved to be one of the best Ferri ever made—until Sycamore, that is. "We have relied on what we learned with the Cascade experience to build up a portfolio of networking companies," Ferri later said.[6] In many ways the Matrix portfolio was incestuous: Cascade's alumni—and there are many—have all gone straight to Matrix's offices when they wanted to start a company. Copper Mountain Networks, Sonus Networks, Sycamore Networks, NetCore Systems, and ArrowPoint were some of Matrix's big wins during the optical boom.

Typically, venture capital (VC) funds raise money from limited partners like pension funds and university endowments, and they depend on their track record to stay in business. As long as VC firms are making money, the limited partners turn a blind eye to everything, including the whopping 20 percent of the profits and 2 to 3 percent management fees that venture capitalists skim off the top. Matrix had it even better—most of its investments were proving to be either home runs or doubles at the very least. Through the 1990s, Matrix, every year, turned each dollar into two dollars for its limited partners, according to *Forbes*. When Cascade co-founder Wu-Fu Chen left to start Arris Networks, Matrix was quick to invest in that maker of networking hardware. In April 1996, about nine months after Arris was bought by Cascade for $177 million, Matrix reaped big profits. Sycamore beat even that when it went public.

Network Makers to Net-Worth Players

Towards the late 1990s, initial public offerings had much better odds than Las Vegas casinos. Investors, big and small, began treating the stock market like a slot machine—put some coins in, and wait for the money to start rolling in.

From 1998 through 2000, IPO shares—whether they were dot-coms, ephemeral business-to-business plays, or super-complicated broadband companies—would rocket on the first day, in part because small investors decided to stampede into these stocks. For almost a century, stocks were the preserve of the well-heeled. But the Internet brought stock prices right to an investor's screen, and buying or selling was a mouse-click away. The democratization of the markets was a welcome change, but there was a down side to it.

"The burst in stock prices over the past four years has coincided with an explosion in investment technology. People look from the Pentiums in their laps to the fat balances in their brokerage statements and naturally make a connection. They think they're on to something," wrote *Forbes* in its cover story, "Amateur Hour on Wall Street." The magazine cited the Institute of Psychology and Markets of Jersey City, New Jersey, when it said that "the average investor expects an annual 18.6 percent rate of return on stock and stock fund portfolios. With dividend yields where they are now, that kind of return would put the Dow at 45,000 in a decade and 210,000 in two decades. A little historical context: Just before the crash of 1929, investors expected an annual return of 10 percent."[7]

Dot-Com Mania to Broadband Madness

In 1999, investors who had gotten fat on dot-com stock offerings decided to shift trillions of dollars to broadband companies, as attention shifted to the companies that provided plumbing for the Internet. By 1999, many investors were finding it difficult to buy dot-com stocks, and infrastructure plays were still relatively cheap. Of course, Wall Street was doing a great job of selling this to the investors—the fast growth of Nortel and Lucent was a primary driver of dollars into equipment stocks. Thus began another hyperinflated bull market.

IPO and the Nine Nines

On January 28, 1999, Pleasant Hill, California–based Tut Systems raised $45 million by selling 2.5 million shares at $18 a share. On its first day of trading, the stock rose 156 percent, to close at $57.50 a share. The company, which made equipment that could deliver broadband into high-rise apartment buildings, was a money loser, with $27.8 million in sales and a loss of $11.2 million. That didn't deter anyone from driving the stock higher. Investors had valued $1 of Tut sales at $33, proving the age-old axiom wrong—money does grow on trees.

On April 9, Extreme Networks of Santa Clara, California, raised $119 million by selling 7 million shares at $17 a share, and the stock gained 226 percent to close at $55.38 (split-adjusted: $27.69) a share. A month later, on May 18, 1999, Redback Networks, a Sunnyvale, California–based company that made equipment for high-speed Internet access, went public—Morgan Stanley Dean Witter helped sell 2.5 million of its shares at $23 a share, and the stock closed its first day at $85, which put the total value of the company at $1.77 billion.[8]

That day, Roger McNamee, an influential investor and venture capitalist who ran Integral Capital Partners in Palo Alto, California, told CNNfn, the financial news channel: "And I think that for the first time in a long time it's safe to go back in the water with data networking stocks that are kind of the new and emerging players because there is really some exciting stuff going on as we put broadband into the Internet." It was as if he was firing the proverbial racing gun for the mad dash into broadband hardware stocks.

In August 1999, Cisco Systems spent nearly $7.4 billion on two tiny optical-networking start-ups, Cerent and Monterey Networks, lending legitimacy to the market's optimism. Cisco paid $6.9 billion for Cerent, roughly 23 times the sales Cerent expected for 2000. The deal would fuel forthcoming public offerings from Foundry Networks and Sycamore Networks, among others.

Extreme Networks, Juniper Networks, RedBack Networks, Copper

Mountain Networks, and Foundry Networks all debuted in the stock market in 1999 and rose to multibillion-dollar valuations, even though together they couldn't muster up sales of even half a billion. In 1999, telecom equipment IPOs averaged returns of 240 percent. These newly public companies were more like net-worth companies than network companies.

These shares also benefited from a Wall Street practice, known as flipping, which works like this: If a company's stock is offered to the public at $15 a share, and it closes the first day of trading at $40, any institutional investor who got 40,000 of these shares could sell them by the end of that day and take home a profit of a cool million bucks. Now if you do this every second day—well, you get the picture. This slice of the pie was, of course, reserved for the very elite—the bankers, the fund managers, the friends and family, and the chief executives of other companies. This was a favorite tool Salomon Smith Barney used to keep its big customers happy.

"From late 1998 through the summer of 2000, Wall Street was an out-of-control beast, and IPOs were its sustenance," *Red Herring* would later write.[9] Investment banks that underwrote the hot public offerings controlled the shares and doled them out to big investors who traded with their firm, or to technology executives who could bring some business to their firm.

Jay Ritter, a professor of finance at the University of Florida in Gainesville, Florida, and a well-respected IPO expert, estimated that the "flipping helped the average 1999 IPO spike 70 percent on its first day of trading and the average 2000 IPO rise 55 percent. That's more than double the average first-day jump of 20 percent in 1998."[10]

But the gains of institutional investors were nothing compared to those of the entrepreneurs who started these equipment companies. Many of the founders and co-founders of network equipment companies became the newest members of Forbes' 400 Richest Americans list. Juniper Networks' co-founders Pradeep Sindhu and Scott Kirens, Foundry Networks' Bobby Johnson, and the Sycamore duo were each worth more than $2 billion. "The telecom industry surely creates more millionaires per day than any other industry," wrote Daniel Briere and Christine Heckart in *Network World*, a trade publication for the data-networking crowd. "Anybody with a glimmer of an idea can start a company and sell it in a couple of years for at least a few hundred million dollars. So why not get a piece of the action?" they wrote.[11]

Paulie Speaks Out

Pundits began talking of how these newcomers could trump Cisco, Lucent, and Nortel. Paul Johnson, an analyst for BancBoston Robert son Stephens, a San Francisco investment bank that is now out of business, was one of the biggest cheerleaders for everything optical. In August 1999, Johnson addressed an audience of more than 200 institutional investors at the investment bank's annual "Investing in Innovations" conference in San Francisco and waxed eloquent about the beautiful and profitable world of broadband stocks. In his presentation, Johnson said, "We estimate that the next-generation network market will, within the next 10 years, see $500 billion to $1 trillion of wealth creation in the next-generation equipment vendors. We expect that the next-generation network market will generate greater wealth creation than any prior technology market we have seen in the past. We believe that no prior leader will emerge as the winner in this space."[12] His favorite companies included Juniper, Redback, and Sycamore Networks.

Johnson's cheerleading has landed him in the soup, and the Securities and Exchange Commission is currently investigating the former analyst. On January 9, 2002, the SEC filed a civil action suit "for issuing fraudulent research reports."

Among the charges: that Johnson issued research reports and made public statements regarding mergers proposed by two public companies in which he failed to disclose that he had conflicts of interest because he owned stock that, upon completion of each of the mergers, would yield enormous financial windfalls for Johnson. The two companies in question: Redback Networks and Sycamore Networks.

Johnson had invested $50,000 of his own personal funds in Siara, purchasing 26,595 shares of Siara's Series B preferred stock, and scored a major windfall when Redback bought that company for $4.3 billion. His take—$9.9 million. And that was not all. He had invested $75,000 in Sirocco Networks on January 10, 2000, and received 23,235 shares of Series D preferred stock in Sirocco. But when Sycamore bought Sirocco for $2.8 billion in June 2000, Johnson hit the jackpot one more time. His total investment was worth $2.3 million by the time the merger closed in September 2000.[13]

The SEC notes in its complaint that during the top of the broadband bubble, in 1999 and 2000, "Johnson praised both mergers in his research reports and media statements, but failed to disclose that his supposedly objective advice was infected by serious conflicts of interest. He did not disclose his personal holdings in the affected companies or the magnitude of his financial interest in the outcome of the mergers."[14]

Like other bubble era analysts, Henry Blodgett of Merrill Lynch and Jack Grubman of Salomon Smith Barney, Johnson also had differing views at the same time on various companies—a negative, more bearish view for his colleagues and another, more sunny view for the investors, the SEC charges.[15] "In 2001, Johnson issued false and misleading buy recommendations on another public company that were inconsistent with his privately-held belief," the commission charged in a press release accompanying the civil action.[16]

The company in question: David Huber's Corvis Corporation. The SEC charges that Johnson told Robertson Stephens' investment committee that he would "not buy Corvis stock at the prevailing market price, but would buy at a price that was approximately half of the current price." However, in his research reports he rated the stock a buy. In addition, the day after he made his private recommendation to the committee, Johnson sold nearly all of his Corvis stock. Two days after his stock sale, Johnson issued another research report reiterating his buy recommendation on Corvis, but failed to disclose that he had sold his Corvis stock two days earlier.

In the bubble years, however, these activities went unnoticed.

The Customer Is Always Right

Sycamore was focusing its energies on emerging carriers such as Williams Communications, and the company customized its products according to the needs of their potential customers. The Lucents and Nortels of the world were selling one-size-fits-all solutions. Sycamore got detailed product specifications from these small players and, within 18 months of starting the company, had developed and shipped a product. That was all it needed to go public. On August 9, 1999, Sycamore filed a prospectus for an IPO with the U.S. Securities and Exchange Commission. It had little or no revenues, and its only customers were still testing the product, but it was shaping up to be one of the hottest networking IPOs.

While I was following the forthcoming offering (which was underwritten by Morgan Stanley Dean Witter) for Forbes, the constant refrain from most institutional investors was that they couldn't get enough shares. Adding to the IPO's appeal were rumors that Sycamore was close to signing Williams Communications as its first customer and, along with that, would announce an order for $24.5 million worth of equipment. (The deal was not announced until after the public offering.)

"On day one this [stock] could trade at over $100, and this offering could easily be bigger than Foundry Networks," said Raj Srikant, then an optical networking analyst with FAC Equities in New York.[17] He was right—the offering, which initially was supposed to be priced between $18 and $20 a share, was eventually priced at $38 per share. Sycamore's original target was to raise $115 million by selling 6.5 million shares, but demand was so high that the number of shares for sale increased to 7.5 million shares. Instead of $115 million, the company raised $284.3 million in its IPO. With analysts like Paul Johnson touting the company to anyone who would listen, it was no surprise that the Sycamore offering blasted out of the gates.

On October 22, 1999, the stock opened at $213.46 a share (not split-adjusted), more than 462 percent higher than the actual price of $38. A few minutes after Sycamore's ticker symbol, SCMR, scrolled across the NASDAQ tape, Desh was worth $4.4 billion, and Smith was worth $4 million. Matrix's 16 percent of the company was worth around $2.8 billion. The Dan and Desh Show was a hit. "Wall Street has been very supportive and has bought our long-term vision. Now we have to work hard to deliver on the promise. Only in this country can an idea and a business plan help you raise millions and become this successful," said Deshpande on the day of the IPO.[18] Investors just bought and bought and bought . . .

"All of us were just standing there and waiting for the stock to open," recalled a former employee of Sycamore. And when it did, "we were all millionaires at the time," he said. There was wild cheering for a few seconds, and then people drank champagne in cheap plastic glasses and went back to work. Of course, Desh and Dan were busy talking to the media, which had been hounding the company for almost a week.

At the Next Generation Networking Show in Washington, D.C., in November 1999, everyone was talking about one thing and one thing only: the Sycamore initial public offering. "Infrastructure is a great place to invest. $200 billion in circuit-switched business is migrating to the packet-switched world," remarked James Wei, president of Worldview Technology Partners, at a venture capital panel.[19]

Sycamore's public offering was the culmination of one of the best years ever for investment banks, especially those underwriting the new shares of companies making optical and data networking equipment. The class of 1999 ended the year with logic-defying triple-digit gains— Juniper finished 900 percent higher and Redback ended the year 1,443.5 percent higher. Sycamore rang in 2000 with a gain of 711 percent. "Their nosebleed valuations leave room for nothing but absolutely heroic outcomes, which are far from assured. After all, there are more than 100 of them; only a handful can succeed," cautioned *Fortune* magazine, but of course no one paid any attention.[20]

Pay to Play

Soon after Sycamore's public offering, in November 1999, the company got a $400 million order from Williams Communications, which added further zip to Sycamore stock. Getting the Williams contract was important for Sycamore to maintain its status as a hot stock. How did Sycamore secure a contract with Williams Communications? Ryker Young, vice president of sales at Sycamore, had an old buddy at Williams Communications: its chief technology officer Matt Bross. Young had been trying hard to convince Bross to buy some of Sycamore's gear, and he succeeded. In September 1999, when Bross was offered Sycamore's hot IPO shares in consideration for business that the company had already won, everything suddenly fell into place. Bross had pioneered yet another way to get rich—the pay-to-play strategy. Actually, it was genius. Bross was like the bouncer at a hot nightclub. The optical start-ups were single guys waiting outside the club, desperate to get inside. A little *baksheesh* usually does the trick outside a nightclub, and this broadband club was no different.

Sean Doherty, a principal at the Palo Alto, California–based restructuring firm the Venture Asset Group and a longtime veteran of the broadband business, said if "the companies did not play the game," they would lose out. Doherty was reminiscing about his days as the chief executive of a now-defunct local exchange carrier.

Matthew Bross had a fairly simple life before he joined Williams Companies' broadband division, Williams Communications Group. The 41-year-old networking expert had joined WCG in 1997 after WCG had bought an Internet company, Critical Technologies, a company Bross had co-founded. After joining Williams, Bross was made the chief technology officer and vice president and general manager of emerging markets. He soon began to literally throw his weight around. In other words, he was the guy deciding which company's equipment would be used in the new network Williams was building. With millions to spend, he now had a stature to match his girth.

A look at Williams' SEC filings shows that Bross didn't make that much in terms of salary—between $275,000 and $350,000 from 1999 through 2001. Nice salary, but he was no millionaire. He got a few thousand stock options from Williams, but that didn't amount to much. That was chump change compared to what some of the other technology executives were making at the time. At a time when he was deciding the fate of companies (by buying their equipment), Bross knew he was making the founders of these companies billionaires. The unfairness of it all must have bothered him, until the realization that he had the power to make or break companies—and therefore to get stock and options from these companies!

And it was all legal—well, almost. According to *Fortune* magazine, "at least a half-dozen Williams employees were given options for Sycamore stock, options that wound up being worth a lot of money."[21] Bross, who got friends-and-family shares, saw his "investment" grow to a million dollars, while Wayne Price, another Williams vice president, got friends-and-family shares that, at the top of the market, were worth $500,000. Price actually did even better—since he was on Sycamore's technical advisory board, he got additional options worth almost $10 million, *Fortune* reported.[22]

While there is legally nothing wrong with this practice, it does blur ethical boundaries. But the laissez-faire attitude was tolerated. Timothy

Barrows, a general partner with Sycamore's early backer, Matrix, even justified the reward in an interview with *Fortune*. "It's a way to give a small hit to the people who've helped with the company. The goal is to spread it around not just to customers that you've booked, but to customers that you want to book, to concentrate the ownership as much as possible in the hands of people who matter. It's free shares. You know they're going to trade up, and you take a small slice of the offering and direct it to anyone who can be helpful in the company," he told the magazine.[23]

Matt Bross was becoming a master at getting shares in the yet-to-go-public companies he signed contracts with. He did an encore with ONI Systems,[24] another optical networking vendor that named him to its board of directors. Then he purchased 322,000 shares of the company. John Bumgarner, president of Williams' strategic investments unit, bought 63,316 shares for $6.32 each, according to the *Wall Street Journal*. The company invested about $10 million in ONI, and then gave the San Jose, California, start-up a contract for $30 million. In other words, the senior management was not only blessing Bross's behavior but had also gotten into the action themselves.

At its peak, on June 22, 2000, ONI was trading at $136.75 a share, and that valued Williams' stake at over $180 million, six times the value of the contract. The conflicts of interest are quite clear, even though during the boom this was a standard practice. Bross's $1 million investment grew to $37 million, while Bumgarner's $400,000 investment was worth more than $7 million. Apparently, Bross was reined in after Williams' head honchos figured out that he was pulling in more money than the carrier's chief executive officer.

Bross was not the only person to profit. Cosine Communications of Redwood City, California, gave away about $17.5 million in warrants to its customers, such as Qwest and the now-defunct Broadband Office. These customers generated about $3.5 million in sales for Cosine at the time the company went public. Hilary Mine, an analyst for Probe Research, told the *Industry Standard*, "This is not a new game. What's new is the disturbing degree to which individuals are getting rich. In the past, there were smaller and subtler ways to strike deals on the golf course. Now there's the big bang of an IPO."[25]

But back to Sycamore . . .

Split and Grow

On January 26, 2000, with its stock trading at $300.06 (pre-split) a share, Sycamore announced a three-for-one stock split. The idea of stock splits is that when the stock gets too expensive, it is hard for small investors to buy into the company. So by splitting, say, a $150 share into three $50 shares, a company can hope to attract more retail customers without actually changing its market value. When Sycamore hit an all-time high of $189.94 a share (that would have been $569.92 before the split) on March 2, 2000, the company sold 10.2 million shares in a secondary offering, raising $1.5 billion.

After that, Desh and Smith were worth around $8 billion each. Several executives sold shares. Among the sellers were most members of senior management. Desh and Smith sold 240,000 shares, raking in about $25 million each. Sales maestro Ryker Young raked in about $30 million. Eric Swanson sold 281,000 and took home about $30 million, and Rick Barry sold 553,000 shares, bringing in $56 million. Chikong Shue, an old compadre of Smith and Deshpande from their Cascade days and Sycamore's vice president of engineering, sold 867,000 shares and took home $90 million.

In June 2000, Sycamore bought a hot metro start-up called Sirocco Systems for $3 billion in stock. The company was also signing new customers, such as Storm Telecommunications, Global NAP, Vodafone, Core Express, and Enron Broadband Services, almost on a monthly basis. But when, on July 7, 2000, the company announced a $420 million contract from 360networks, the stock markets went into a tizzy, sending the stock up $16.75 a share to split-adjusted $126.94 (or $378 a share without adjusting for splits). For a while, it seemed Sycamore could do no wrong. Sales for fiscal 2000 had boomed to $198 million from $11.3 million a year earlier.

Through 2000, overoptimistic forecasts from research organizations like Ryan Hankin and Kent (RHK) of San Francisco continued to help move hot money from dot-com stocks to the broadband sector. In January 2000, the firm released a heavily hyped Internet traffic report. "Bandwidth requirements at all points in the network will continue to increase," RHK noted in a press release.[26] "Service providers and carriers will need to deploy more systems with higher capacities to support this continuing growth of Internet traffic," said RHK's director of edge switching and routing, Tracey Vanik.[27] In other words, it was all right to buy the broadband stocks, especially those of equipment makers, because their sales were about to explode.

Then, in October 2000, RHK predicted that global capital expenditures by service providers would grow from $306 billion in 2000 to $370 billion in 2001, a 21 percent increase year-over-year. Dr. John Soden, managing director of RHK's Financial Advisory Services, noted, "These figures indicate continuing massive growth in spending for optical-networking equipment, albeit at a slower year-over-year growth rate than the torrid pace of 2000."[28] There was an absolute disregard for reality, it seems. With dot-coms, one of the biggest consumers of bandwidth and hosting operations, closing at an alarming rate, and with increased bandwidth capacity coming online, it was clear that the bandwidth business was about to nose-dive. There was soon going to be a supply-demand imbalance. That would ultimately have an adverse impact on the total demand for new equipment, but was anyone thinking that far ahead?

"There was a feeling that we could charter our own destiny. We were signing contracts with everyone and a certain amount of arrogance came into the company," a former Sycamore employee said. To some, Sycamore's customer wins were mystifying. In October 2000, Lucent and Nortel said there was a slowdown in optical demand, and they reconfirmed that in early 2001. In the carrier markets, the price of bandwidth was in a free-fall, and bankruptcies were looming large. These were ominous signs for the whole industry, but the Sycamore juggernaut rolled on, at least until April 2001. When the slowdown in demand finally trickled down to Sycamore, the company had to tell investors the bad news, and the company's Cinderella story came to an end very quickly. Since many of its customers were filing for Chapter 11, the company's sales fell off the cliff. The company told investors that the sales for fiscal third quarter would be between $50 million and $60 million, down from second fiscal quarter 2001 sales of $149 million. The telecom meltdown had finally caught up with Sycamore.

The first of Sycamore's customers to file for Chapter 11 bankruptcy protection was 360networks. Then came Enron Broadband, Core Express, and Williams. "It was just heartbreaking to just hear the news, one by one," recalled a former executive with the company. "We had so much tied to that 360networks deal." The company morale slipped bit by bit.

By October 2001, layoffs were looming large. Sales, which had peaked in the quarter ending January 27, 2001, at about $149 million, were down to $21.2 million, off by 86 percent. The stock, which at one time was close to $320 a share, had slid into single digits.

But by then insiders had sold $726 million in stock. Deshpande sold $137 million worth of stock, while Smith and Shue took home $129 million and $122 million, respectively.[29] Both Smith and Desh, generally decent men, have given away substantial portions of their money to charity. But that is still little consolation to the small investors who piled their money into the stock and lost it.

To be fair, Sycamore did have its real successes: Up until October 2001, the company sold $600 million in gear, and for a while it was profitable. So what happened? The company became arrogant and failed to consider the possibility that things could go wrong. Its products were catered to new-age carriers like 360networks and Williams Communications, who, like Sycamore, had a lot riding on new technology and new business models. Sycamore did not stop to consider that an oversupply situation could arise, or that its highly leveraged customers could go bankrupt. When these carriers collapsed, Sycamore's orders collapsed, too. Dan and Desh hadn't been farsighted enough.

"And when Ryker left, it was over," said a former insider. Some believe that the buzz around companies like Qwest, 360, ICG Communications, and hundreds of others was so high that it was easy to believe that the new carriers would become bigger than AT&T. "I know Desh and Dan reasonably well and I think they would argue that at the time the perception was that [next-generation carriers] were going to take over the world," said Chris Noel, a former colleague from Cascade. "I think they believed that they were successful at Cascade, and they would take over the long-haul business market, but that did not happen," said one industry veteran. The early success and hype was the reason Desh and Dan took their eyes off the ball, and as a result missed the signs of a coming meltdown.

"[Optical networking] was a great concept, but maybe too great. Everyone got into the game and then came the overinvestment. The trouble was not that the concept wasn't right. The trouble is, in a capitalist society, there is no way to control who can invest in what," said Deshpande, with the wisdom of hindsight.[30] In an interview in October 2002,

Deshpande told me that two years previously, no one had the time to develop a coherent strategy because "you were running to meet deadlines and grow." But "it is different right now, now we have a strategy," he said. Like some of its peers, such as Juniper and Ciena, the company is fighting to stay alive. In the last 12 months, Deshpande and his team have tried to develop products that incumbent carriers like Baby Bells can use. BellSouth, a Baby Bell, is buying their gear.

But it is going to be a long haul. For the fourth quarter ending July 31, 2002, Sycamore said its loss widened to $73.5 million on lower sales and higher expenses for job cuts and closing offices. Sales had dropped 83 percent to $8.55 million, and they will go lower before the company starts planning a comeback. Sadly there will be no second chances for those who bet on the company's stock. Sycamore now trades for $3.22 a share.

12

JUST AN ILLUSION

After living in the United States for 16 years, venture capitalist Vinod Khosla moved his family back to India in 1992, and for three years worked part-time for venture firm Kleiner Perkins Caufield & Byers. By 1995, correspondence by e-mail had become de rigueur for real techies, and Khosla, who was one, tried to communicate by e-mail with his KPCB colleagues. But it was an extremely tiresome process. "In India I was trying to send e-mail through IIT (Indian Institute of Technology) Delhi's Education Research Net, and dialing into it was so difficult that it was then I realized the importance of communications," he recalled. Khosla saw the need for big fat pipes to carry data, since he anticipated that Internet traffic would grow so rapidly that it would easily overwhelm telephone companies' networks. In 1995, when he returned to the United States and to KPCB full-time, that thought was firmly entrenched in his mind.

At that time, when everyone—including Khosla's equally illustrious and famous partner, John Doerr—was obsessing about dot-com companies, Khosla drew up a game plan to invest in companies that would build equipment needed to support billions of Internet users. Ever since then, Khosla has brought his big idea to life in a very calculated manner and has stayed 24 months or so ahead of the market. For a brief while, when it came to broadband, Khosla's austere offices on Sand Hill Road in Palo Alto, in the heart of Silicon Valley, were the center of gravity.

By the summer of 1999, Khosla had already had his share of hits. He had invested $3 million in the Web portal Excite, which was acquired by @Home Corporation for $6.7 billion, and his $3 million investment in the networking company Juniper Networks had increased in value by more than 66,000 percent. Plus, there was that little company Khosla had co-founded in 1982—Sun Microsystems.

But it was with Khosla's investment in Petaluma, California–based Cerent Corporation—a small maker of optical networking equipment—that he would make his mark on the tech world. In August 1999, Silicon Valley networking giant Cisco Systems bought Cerent for a whopping $6.9 billion, and Khosla's $8 million investment was suddenly worth $2.4 billion. Three months later, the network equipment maker Redback Networks acquired Siara Systems, another Khosla investment that shared DNA with Cerent, for $4.3 billion, despite not having a cent of revenue to its name.

Even on Silicon Valley's Sand Hill Road, which is peppered with venture capital firms the way urban downtowns are dotted with Starbucks cafes, the Cerent deal was impressive; it was rare to see such a small and closely held company achieve such a valuation. The 46-year-old native of Poona, India, was having one heck of a year.

It's said that when Khosla installed a satellite dish at his Portola Valley, California, home to get high-speed Internet access (because DSL service wasn't available in the neighborhood), others raced to copy his move. Similarly, Khosla's investments in the optical networking world would be widely copied. The sales of Cerent and Siara, along with the spectacular initial public offerings of Sycamore Networks, Redback Networks, and Juniper Networks in 1999, brought a certain madness to the staid venture capital world. The optical equipment bubble, not the last but surely the most damaging chapter of the broadband saga, was born.

Singh Side Story

Khosla's role in these start-ups is akin to that of a championship football coach. But coaches need to be able to draft star quarterbacks to translate their ideas into reality—as did Khosla. His quarterback was Raj Singh, a fellow Indian, who in January 1996 stopped by KPCB's offices looking for help. Singh, an engineer and serial entrepreneur, would go on to found Fiberlane Communications, a start-up that later morphed into both Cerent and Siara Systems.

Singh was born on July 1, 1946, in the small north Indian farming village of Idrishpur, in the state of Uttar Pradesh. This is not the kind of farming community you'd find in the American Midwest. Instead, it is a desolate place where farmers struggle to eke out a living using antiquated tools. Studying under earthen lamps lit by kerosene oil, Singh managed to rise above his circumstances and graduated from the University of Roorkee with a bachelor's degree in electrical engineering. After graduating, he held jobs with the Indian Navy and in Libya with the Libyan Electricity Board.

In 1979, at the age of 34, Singh emigrated to the United States to attend the University of Minnesota for graduate studies. He was then a father of three and older than most of the other students in the program. "It was cold, very cold, and I had never been this cold before," is all he said about those days of hardship. After finishing his graduate degree, he moved to Silicon Valley, and worked at companies like National Semiconductor, Trilogy Software, Cirrus Logic, and NexGen. (NexGen would later be sold to Advanced Micro Devices, and its chips would help AMD register key victories against Intel in the PC chip wars.)

While at NexGen, Singh was bitten by the start-up bug. "A colleague of mine, Eli Sternheim, was talking to me about how difficult it was to learn Verilog HDL (a chip design language) from its manuals, and we both decided to write a book on this subject," recalled Singh. The resulting book, *Digital Design and Synthesis with Verilog HDL*, was self-published and turned out to be a hit. Using the proceeds from the book, Singh, in 1991, started InterHDL, a company that developed verilog-based chip design tools. In 1998, Avanti bought the company for $35 million.

Singh then started another company, Advance-Cel, which developed specialized chips for use in phone networks that used a kind of technology called SONET, which stands for Synchronous Optical NETwork, a standard for transmitting data over optical fiber lines. Instead of making the chips itself, the company licensed designs for them to other major chip players like Adaptec, IDT, and Cypress Semiconductor.

In January 1996, Singh had a new idea for a company that developed Java processors. Developed by Sun Microsystems, Java was a new kind of computer language that allowed programs written in it to run on any kind of computer—Macintosh, PC, or UNIX. Java's unique qualities also made programs written in the early versions of the language painfully slow, and Singh figured a special chip could help boost their performance. Kleiner Perkins had started a special Java fund and was looking for companies to fund at the time, so Singh stopped by the KPCB office with his business plan.

The business plan was intended for John Doerr, who at the time was managing Kleiner Perkins' Java Fund. But it ended up with Doerr's partner Vinod Khosla, who was lukewarm to the idea. Singh knew Khosla from earlier—Khosla had been one of NexGen's biggest champions and was a financial backer of the company. Also, Juniper Networks, another of Khosla's portfolio companies, had licensed technology from Singh's Advance-Cel.

"Mr. Khosla told me there was no money to be made in Java, but we talked about doing a [optical] hardware box which could basically do SONET over optical networks," recalled Singh. Khosla's view was that the sharp increase in Internet traffic would create a market for a device that could handle large amounts of voice and data.

For months, the venture capitalist and the entrepreneur talked about a box that could combine two different types of technologies—Internet protocol and SONET—in one box. The idea was to develop a device that would allow phone companies to offer Internet and traditional voice services cheaply, by using the power of optical networking technology. Starting a network equipment maker was a radical idea, because most phone companies bought their equipment from Nortel, Lucent, or Alcatel. Lucent, especially, had a sales advantage, since the company was spun off from AT&T. But Khosla and Singh believed that price-performance benefits would force phone companies to look at this new box.

Khosla gave Singh $250,000 to start Fiberlane Communications. Then, a Hewlett-Packard salesman and a friend of Singh's introduced him to Ajaib Bhadare, a senior director of engineering at DSC Communications, a phone equipment company. Singh met the British-born Bhadare at an Applebee's restaurant in Petaluma, California, in November 1996.

"They said, 'We've got this market opportunity. Vinod is behind us. But we don't know what the product is or what the architecture would be,' " Bhadare later told The Santa Rosa *Press Democrat*.[1] "I said, 'I'm not sure I want to join you guys. You are strangers. But, if I was to put something together, here is what I would present.' " As they talked, Bhadare took a cocktail napkin and quickly drew the rudimentary design for what eventually became a Cerent box. He left the meeting without any intention of joining Fiberlane.

Bhadare did not want to relocate to San Jose from his home in Petaluma, 83 miles north of Silicon Valley. In order to get him to join Fiberlane, Singh agreed to have a satellite office in Petaluma. Two other DSC employees, Paul Elliott and Hui Liu, also signed on to Fiberlane. Then Singh recruited Al-Noor Shivji and six others from telecom equipment maker MPR-Teletec in Vancouver, British Columbia, brought them to Silicon Valley, and opened an office in San Jose, California. Soon, the company gathered momentum, but in the summer of 1997, disaster struck. The Petaluma engineers and the San Jose engineers began griping about the direction of the company. The Petaluma gang, led by Bhadare and Mike Hatfield (who had joined Fiberlane from Advanced Fibre Communications, another Petaluma telecom equipment maker in April 1997), wanted to focus on voice technology, while the Silicon Valley engineers wanted to go after the data market. There was also underlying tension as many in San Jose resented the Petaluma gang's independence.

In September 1997, the bickering reached a fever pitch. No one was talking to each other in the company. "There were many times when we almost felt like throwing in the towel," said Bhadare.[2] The company was losing some of its momentum as Nortel and, to some extent, Lucent started boasting about their optical products. Meanwhile, Ciena Corporation, one of the first optical start-ups, had gone public in 1997 and had received a rousing reception from investors. The demand for optical gear prompted other companies to jump into the fray, perhaps sensing that big demand was coming from new carriers like Qwest and Level 3.

Fiberlane needed to shift gears, but it couldn't get out of neutral. By 1998, the situation had become so bad that the board of directors decided to split the company into two. "I think the whole split was prompted by some people's desire to pursue different markets and the location of the two divisions," recalled Singh, who by then had left the company to start Stratum One, a communications chip company.

"We ended up with two locations. That ended up polarizing the company and the two teams didn't get along. And too many people are driven by who they like and don't like," said Khosla. "We [had] hired Carl Russo [as CEO] and I said his first call was to decide whether to keep it as one company or split it. At first he decided to keep it together and have one company. [But] then enough people in Petaluma complained about it, [and] he said it's not going to work."

Russo's decision not to split the company in two led to the resignation of 15 people, including Mike Hatfield, who had briefly been chief operating officer of the company. The acrimonious battle finally ended in May 1998, with the 70-employee Petaluma operation changing its name to Cerent Corporation, with Russo as the boss, and the rest of the operation in Silicon Valley being reorganized as Siara Systems. Khosla invested in Siara as well. All through the process of the company's split, Cerent's engineers kept working, and their focus helped the company develop the box quickly. By early 1999, Cerent was ready to show its wares and go after customers. On July 22, 1999, the company filed a prospectus with the SEC for an initial public offering that never happened. It would be acquired in August 1999, in a deal that would send shockwaves through the industry.

Cisco's Quandary

In early 1999, as data and optical networking start-ups began to sprout like weeds in Silicon Valley, John Chambers, the chief executive officer of data networking equipment behemoth Cisco Systems, was worried. Cisco is to Silicon Valley what General Electric is to industrial America. Cisco was a company that pioneered a new market, sold billions of dollars of equipment, and made thousands of its employees into millionaires. It was one of the early successes of Silicon Valley, following in the footsteps of Intel Corporation and other chip companies.

Cisco was started in 1984 by Leonard Bosack and his wife Sandy Lerner, a couple who worked for Stanford University in Palo Alto, California. Bosack was in charge of the Computer Science department's computers, while Lerner was managing computers for the Graduate School of Business. In 1981, the University had received new Alto Workstations and Ethernet boards from Xerox Palo Alto Research Center (PARC). (Ethernet was a technology that enabled computers to network, and it was developed by Bob Metcalfe, one of PARC's residents.) The university staff worked hard to come up with ways to network the new workstations over large distances. While it was possible to connect computers in close proximity to one another, it was difficult to connect them over large distances—like, say, across campus.

The very smart bunch of people at the university, including Andy Bechtolsheim—who later started Sun Microsystems with Vinod Khosla, Bill Joy, and Scott McNealy—came up with a small device they called the "blue box." Stanford's medical school research engineer William Yeager wrote the code for this box.[3] The box made it possible to network computers spread across long distances. Think of the blue box as an extension cord for the network. Every time you need to go farther, plunk one in, and go on your merry way. This was the prehistoric version of a standalone router, a device that is to today's Internet what the microprocessor was to PCs.

A few months later, in 1985, Bosack and a colleague, Kirk Lougheed, asked Yeager for his original software. They later modified it to handle only Internet protocol traffic.[4]

Bosack had already co-founded Cisco by then, hoping to commercialize the blue boxes. He asked Stanford to allow him to sell the blue boxes commercially. Stanford refused, so Bosack and his wife started building their own version of the blue boxes out of their house in Atherton, near Palo Alto. With the basic science of the boxes done, it did not take the husband-and-wife team long to start selling the boxes. All this time, Bosack was still working for Stanford and was using the university as a calling card to sell Cisco gear.

In 1986, university authorities asked Bosack to choose between Stanford and Cisco. On July 11, Bosack left the university along with his old friend Lougheed as well as Greg Satz, a programmer, and Richard Troiano, who later handled Cisco's sales. Stanford and Cisco continued to fight over intellectual property rights, and legal threats were brandished about. Finally, on April 15, 1987, Stanford licensed router software and two computer boards to Cisco, and in exchange received about $170,000 (including royalties). The deans of academia may have cut themselves a poor deal.

One of the keys to Cisco's success was that its products could connect computers across long distances, regardless of who made the computers. This was unusual in a world where few computer technologies were compatible. Cisco's products could make IBM machines connect with computers made by Digital Equipment Corporation or by Wang Computers, because they used the Internet protocol. In the early days, the company sold the blue box variants to other engineers. But then came a big break.

A big account won from Procter & Gamble helped increase orders for Cisco's blue boxes. "We suspected that Procter & Gamble in Des Moines was going to want to talk to Procter & Gamble in San Francisco," Lerner later told *Forbes*.[5] By the end of 1987, the company was doing a quarter of a million dollars in sales every month, but since it was self-financed it was running out of cash. The more sales, it made, the more raw products it needed to grow, and that was not possible without cash. So it decided to raise venture capital.

Don Valentine, a general partner at the venture capital firm Sequoia Capital and the man who had originally backed Apple Computer, decided to help out. He invested $2.5 million in Cisco, perhaps one of the best investments made by a venture capitalist at any time. But Valentine's investment came at a cost. He brought in professional management in the form of John Morgridge, and Bosack and Lerner lost control of their company. That was when the company started to change. Until then, the company was mostly staffed by Lerner's and Bosack's friends and family. By 1990, the founders and Morgridge were having arguments, and Lerner was fired. Bosack decided to quit.

On February 16, 1990, Cisco went public and quickly became the stock market darling. Its products were moving off the shelves at an amazing clip, thanks to a crackerjack sales force that was highly focused. Many competitors cropped up, but they failed to make an impact on Cisco. In January 1995, Morgridge handed over the reigns to John Chambers, a rising star in the Cisco ranks.

Born in Charleston, West Virginia, Chambers oozes Southern charm and has an accent that puts everyone around him at ease. He is the son of two doctors and overcame mild dyslexia to graduate second in his high school class. He attended West Virginia University and later earned a law degree there. He then received an M.B.A. from Indiana University. Chambers is married to his childhood sweetheart, Elaine Prater, and is the father of John Jr. and Lindsay. His preacher-like tone, handsome looks, and good manners made him perfect for the job of a salesman at International Business Machines, and later at minicomputer pioneer Wang Computers.

In 1991, when Wang was spiraling downwards, Chambers joined Cisco as senior vice president of worldwide sales and operations. He rose through the ranks as his sales prowess and diplomatic demeanor made him the heir apparent at Cisco. In January 1995, when Chambers became chief executive, the company's sales were $454 million, with its stock trading around split-adjusted $2 a share. Cisco's main products, its routers, had become bigger and beefier. But Chambers knew that the world was going to change soon. He embarked on a dangerous, high-stakes strategy: growth through acquisition.

Buy, Grow, Buy Some More

From the time Chambers took the top post, Cisco made, on average, 8 to 10 acquisitions a year and expanded its product offerings at warp speed. Most of Cisco's acquisitions were driven by customer needs. Whenever a customer suggested that it needed a certain product, Cisco went out, and bought the best technology available.

The rising Cisco stock—$23.20 at the end of 1998—was a perfect tool to retain the managers and employees of the acquired companies. This was key, because many acquisitions fail when the brains from the acquired company hit the road. "You couldn't genetically engineer a better leader. Cisco makes every acquisition feel they're part of the company. It represents the best of Silicon Valley culture," Sun co-founder Andy Bechtolsheim, who sold his data networking company Granite Systems to Cisco in 1996, told *Fortune*.[6] The mergers resulted in Cisco growing big really fast. The company's offices now sprawl along South Tasman Drive in San Jose.

Chambers was a popular chief executive. He would serve ice cream to employees and go around introducing himself to them. And he would remember their names years later. By 1998, Cisco sales had touched $8.5 billion,[7] and its market capitalization was about $108.4 billion. The stock had done so well that Cisco's employees were viewed with envy by almost all their peers. Cisco was the coolest company to work for, and everyone wanted to work there. But still, Chambers was worried.

Boom Goes Optical

Even as the Cisco monster grew and grew, there was a perception that the company was missing out on the broadband boom. Most of its products catered to the needs of large- and medium-sized corporations and institutions. It was a major supplier to Internet service providers but had very little presence in the new optical networks. It also needed to get into the broadband access business—which meant selling equipment to DSL service providers and other new phone companies. The back-haul networks, which connect cities to one another, used optical networking gear from upstart Ciena, or from the more established Lucent and Nortel. Cisco didn't sell any optical gear, and it needed to enter this market, according to Bill Lesieur, an analyst with Technology Business Research, a research group based in Hampton, New Hampshire.

The reason for Chambers' worries was purely economic. Sales of optical gear by rivals like Ciena, Lucent, and Nortel to telecom carriers were exploding—they were up to $8.49 billion in 1998. Nortel's sales went up 36 percent for the same time period, while Lucent sales had surged 29 percent. Even a pesky upstart like Ciena had sales growth of 36 percent (between 1997 and 1998). In addition, Lucent, of Murray Hill, New Jersey, and Nortel, of Brampton, Ontario, were both pursuing acquisition strategies and were moving right into Cisco's backyard. Nortel had bought Santa Clara, California–based Bay Networks, and Lucent also had an acquisition plan.

The world of networks was changing quite drastically. For decades, phone companies built their business around circuit-switched networks, but after the Telecommunications Act of 1996, there was a belief that networks based on Internet protocol would carry both data and voice traffic. "Cisco believed that eventually all traffic—voice and data—would travel over the same Internet protocol–based network. Cisco entered the voice networking market from the data networking side, and Nortel and Lucent entered the data networking market from the voice side," said Lesieur. "Cisco was selling the vision that IP would change the world, so you better get on board or you will get wiped out."

Cisco's target market was the emerging carriers, which lacked legacy networks and were more likely to adopt Cisco equipment as well as rely on Cisco as a technical and business advisor. Cisco's vision was that these emerging carriers, all running Cisco equipment, would disrupt the incumbent carriers completely, according to Lesieur. As the emerging carriers won, the thought was that Cisco would take over the service provider market. Cisco was betting that the new IP-based carriers would use its gear and take market share away from legacy phone companies, because it had superior technology. Cisco believed it would dominate the market in the long term.

After a lot of internal debate, mostly about strategy, the company decided to get into the business of selling equipment to service providers. After all, the $225 billion market for telecom gear was at stake. Two key backers of the move were Don Listwin, executive vice president of corporate marketing, who many saw as Chambers' heir apparent, and Kevin Kennedy, senior vice president of Cisco's service provider business. Once he made up his mind, Chambers went full speed ahead. He made his intentions clear at the January 1999 Las Vegas Consumer Electronics Convention in a keynote speech, where he talked about free voice calls, the Internet, and how the Net changed the world of commerce. Cisco launched a $40 million marketing campaign that asked the obvious question: "Are you ready?"

Cerent's Big Payday

Despite all this talk about how Cisco powered the Internet, the company had little or no equipment that supported the new network backbones based on optical technologies. Cisco was like a car company that could make the gearbox, the steering, the body, and the brakes, but not the engine that drove the automobile. The spending just on long-haul optical network systems was around $2 billion in 1998 and was projected to grow 70 percent over the next four years. The total market for optical gear was even bigger. It was simply too good a market for Cisco to ignore. Chambers needed to come up with a winning optical strategy—fast.

In May 1999, at a telecom shindig in Laguna Beach, California, Chambers met Carl Russo, chief executive officer of Cerent. "How much would it cost me to buy you?" he asked Russo. "How much would it cost for you to leave us alone?" replied Russo.[8] With its product already finished and ready to ship to test customers, Russo had a reason to feel bullish about his company's prospects. It had a product that Cisco did not have, the demand was strong, and the company was hoping to go public in the fall of 1999. Cerent's box could help Cisco combine its IP expertise with optical networks.

Still, Chambers would not go away. Ten weeks later, Chambers called on Russo again. This time around Russo agreed, mostly because Cisco offered a price that was more in sync with the valuations of networking start-ups that had then recently gone public. By August 13, 1999, Cerent and Cisco had negotiated a $6.9 billion deal.

Vinod Khosla apparently was vacationing in the Galapagos Islands then when his satellite phone rang. Russo brought him up to speed. Khosla caught the first helicopter out and signed off on the deal. He had turned an $8 million investment into $2.4 billion—not bad for three years of work.

Cisco and Cerent announced the merger on August 26, 1999. It was major news, and even the *Wall Street Journal* was impressed by the sheer amounts involved. "It was the market and the euphoria of the time that when the Cerent deal was announced, I wasn't that shocked because companies were going public for billions of dollars," said Bhadare, who, as one of the founders of Cerent, was suddenly worth $160 million in Cisco stock.

"I was shocked that someone would pay those kinds of valuations. I think at the time of the deal there wasn't anyone in the company who wasn't a millionaire—even my admin [assistant] for a brief while was a millionaire," recalled Bhadare. As for Russo, his share was worth over $300 million. Russo had made Cisco pay top dollar for Cerent.

Cisco typically paid between $60 million and $200 million for a start-up, or roughly between $500,000 and $2 million per employee. For Cerent, Cisco spent $23.3 million per employee. The market applauded Cisco's move and drove up the stock, betting that Cerent would help Cisco get a toehold in the fast-growing telecom market. With a $225 billion market up for grabs, many analysts thought Cisco was getting a bargain. Chambers was the king of Wall Street. On his watch, the company's sales had gone from $1.3 billion in 1994 to almost $12.2 billion in the fiscal year ending July 31, 1999. The stock was up almost 2,300 percent, and the company's market capitalization was $220 billion.

A few months later, in October 1999, Redback Networks decided to buy the other half of Fiberlane, Siara Systems, for a whopping $4.3 billion. Khosla definitely was the new genius in Silicon Valley. *ECompany Now* magazine (now *Business 2.0*) called Khosla the sorcerer and Raj Singh the apprentice. (Singh later founded Stratum One, which he sold to—guess who—Cisco.) *Red Herring* magazine put Khosla on the cover, calling him "The No. 1 VC on the planet."[9] "I had no idea that Fiberlane would become this big, given that we had to go through so much difficulty in the early days," Singh told me later.[10] With about 2.5 to 3 million Cisco shares in his brokerage account, Singh had a $300 million windfall.

"What the Cerent deal signaled was that the companies [like Cisco] were willing to acquire start-ups as private labs, and products. It was a business model which said—grow by buying R&D," said Steve Kraus, a general partner with the Silicon Valley venture group US Venture Partners. And this prompted venture capitalists to fund companies at a rapid clip.

These two seminal deals fueled an optical networking bubble. Any start-up with the words *photonics* or *optical networking* in its business plan could expect to raise millions of dollars in venture capital. Venture-One, a research group, estimated that in 2000, 198 companies raised about $6.8 billion, with average company valuations rising from $5.2 million in 1998 to $30.4 million in 1999. Everyone looked at Kleiner Perkins Caufield & Byers' $8 million investment in Cerent, and its returns, and it was not difficult to hear the *cha-ching!*

"The way the financing is flowing, you would think that tomorrow we are going to see the advent of this new faster, cheaper, more efficient network—but it will take some time," said Robert A. Saunders, senior analyst with The Eastern Management Group, a Bedminster, New Jersey–based telecom consultancy. "I think that we are in the midst of an overexuberant financing of this particular sector," he continued.

It was not just Cerent and Siara. Valuations of public companies were going through the roof. Sycamore Networks had made a spectacular debut in the public markets, and tiny Juniper Networks and others were being valued in billions. For the venture community, it seemed that an investment in an optical networking start-up couldn't go wrong. Either the company could count on an IPO home run or it could potentially be bought by Cisco or another networking company for billions. Since Wall Street was seeing the newcomers post strong revenue growth, and valuing them accordingly, the overexuberance spread to the private sector as well. "Valuations became bizarre," said Kraus. "We had Procket Networks in our portfolio and we got a market valuation for a billion dollars. Agility was valued at $400 million." These were early stage companies.

Venture capitalists were throwing caution to the winds. Between 1996 and 2001, a total of $8.7 billion was invested in different sorts of broadband-related technologies, including optical technologies. But during the same time, Cisco, Lucent, and Nortel spent $105 billion on acquisitions. The returns for gambling on optical start-ups were well worth the risk, if Khosla's track record was any indicator.

Even more staid companies like Alcatel, Marconi, Ciena, and Tellabs were going on a buying binge. Start-ups like Sycamore and Redback Networks were also snapping up companies. Data accumulated by Broadview International, a Silicon Valley M&A advisory firm, showed that between 1996 and 2002 the total mergers and acquisitions in the telecom equipment industry in North America and Europe topped out at $189 billion—with $71.4 billion coming in 1999 alone. (See Figure 12.1.) "I think that post Cerent acquisition, everyone was building their

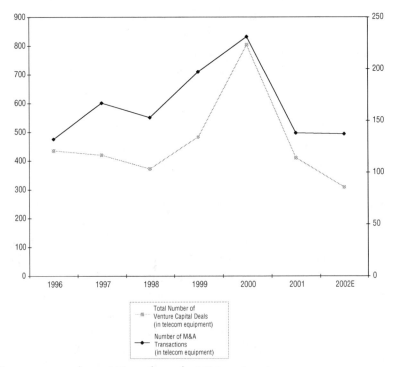

The mergers and acquisitions drove the VC investments . . .

Figure 12.1 Which came first?—VC Bubble or Merger Madness?
Sources: Venture Economics, Venture Expert, and Broadview International

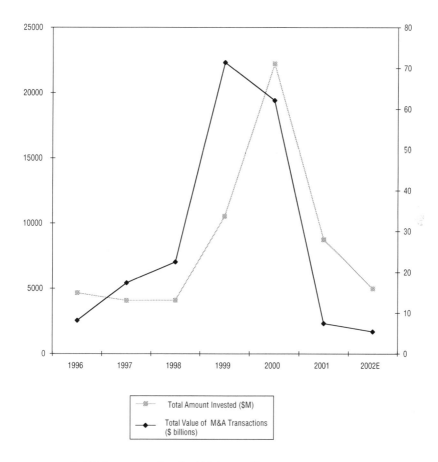

. . . as greedy VC firms ran after the M&A payoff.

Figure 12.1 *(Continued)*

cases on an acquisition as an exit strategy," said Steve Kraus. "It was a to-tally new approach to the world."

Components Are Up Next

In March 2000, when the dot-com bubble burst, investors plowed their dollars into the optical and broadband stocks. Though the NASDAQ dropped 200 points on March 14, 2000, Cisco's stock hit an all-time high of $80.06 two weeks *after* that, on March 27, 2000. Around this time something new happened in Silicon Valley—the arrival of the giga-fund, a venture capital fund with more than $1 billion under management. Several well-known and established venture capital firms like New Enterprise Associates and Redpoint Ventures raised $1 billion–plus funds and threw a lot of those dollars at optical companies. But the timing of the investments could not have been worse. Most of the carriers that bought all the equipment were experiencing problems of their own, and their capital expenditures were being slashed. The venture capital firms didn't factor in the reality that it takes a long time, typically around three years, to develop and market a good product. In other words, these venture capitalists would have to look beyond their first-round investments, and try to ensure that their companies stayed alive and well-funded for the following three years.

Basically, the venture capitalists had gotten in too late. In the euphoria of the moment, many overlooked the fact that the value of acquisitions was declining. Cerent's $6.9 billion price tag was an all-time high, followed by Redback Networks shelling out $4.3 billion for Siara, Lucent buying Chromatis Networks for $4.5 billion, Nortel paying $3.2 billion for Xros and $3 billion for Qtera, and Ciena spending a mere $2.3 billion for Cyras Networks. Still these were big-ticket mergers and were enough to keep the venture capitalists happy.

Worried that the equipment sector might be getting overfunded, the venture capital community turned its attention to optical component makers. JDS Uniphase, one of the top component makers, was gobbling up companies at an astonishing rate. It had bought E-Tek Dynamics and SDL for about $56 billion and was hungry for more. Corning, a New York–based company known more for its cookware than its optical components, had just swallowed a virtual unknown, NetOptix, for $2 billion and was looking for more. When asked about the future of fiber optics, Corning's then CEO Roger Ackerman told me, "Oh, we have just begun! This technology is like a four-month-old baby, with lots of growing ahead."[11] Corning was on a high—it was making fiber, the very raw material that was needed to make everything optical happen. It had decided to get ambitious and spread its tentacles into optical components as well. Investments in the fiber-optic components shot up dramatically to $6.8 billion in 2000.

Corning, JDS Uniphase, Bookham, and several others used their high-flying stocks to acquire anything they could lay their hands on. It was the same old story. Nortel decided to do a public offering for its components business (which never happened), and Alcatel decided to issue a tracking stock for its Optronics business in the fall of 2000. The madness spread to other sectors such as communications chips. Many new companies that made chips for broadband networks were funded between 1998 and 2000. Like every other sector where there was excess venture capital investment, the communication chip sector was propped up by the insatiable appetite of Broadcom, PMC Sierra, Applied Microcircuits, and Intel. This prompted a culture of here-and-now investing, and no one considered even for a minute that the boom could ever come to an end.

"Cisco showed the world how to grow through acquisitions," said Sanjay Subedar, a general partner at Storm Ventures. He sold his earlier effort Stratacom [not to be confused with Raj Singh's Stratum One] to Cisco before the current broadband boom, and then later sold E-Tek Dynamics to JDS Uniphase for $15 billion. "The whole thing was not real—one day we are 400 employees and $70 million in sales, and then two and a half years later we are at $750 million in sales," he said about E-Tek's dynamic growth before it got bought by JDS Uniphase. The general consensus at the time was that bandwidth was like memory—you collect as much as possible and then figure out a way to use it. As a result, every company scaled up its operations—be it Corning, JDS Uniphase, SDL, or E-Tek. Looking at the booming sales of companies like JDS Uniphase prompted a lot of academics to jump into the start-up game. "I think a lot of things which should have [first] been researched in academia became a PowerPoint [presentation] and $30 million in funding," said Subedar.

Huber's Hubris

Many have wondered what the top of the optical market was, and the general consensus is that it was the Corvis initial public offering in July 2000. Corvis was the second company started by David Huber, who had been dubbed the "Light Knight" by *Forbes* magazine.

Born in Oregon to middle-class parents, Huber at age five took apart a washing machine and put it back together again. He continued to do similar feats for the next couple of decades until he found his true love—fiber optics. In 1989, Huber started working for General Instrument Corporation, where he managed the light wave research and development program. It was there that he discovered the magical possibilities of fiber optics. Huber was part of a group that figured out how to slice rays of light into different colors and then use these individual colors to send data from one point to another. Originally, this technology was going to be used to send video signals over cable networks, but in the end GIC shelved the project. Huber left but managed to convince his employer to let him take the technology with him. In 1992, he started Ciena. He got venture capitalists to fund the company, but they did so on one condition—Huber would have to hire a professional CEO. He agreed reluctantly. When Ciena's public offering debuted on February 1997, it was worth $2.1 billion, despite the fact that it had less than $20 million in sales. Huber's cut was over $300 million, which should have been enough to make him happy.

But he wasn't—he wanted to run his own company, and as a result constantly bickered with the management team. He was a control freak, and felt that others like Ciena's then CEO Pat Nettles and venture backers like Jon Bayless of Sevin Rosen Funds had taken the company away from him. "Huber is becoming the Howard Hughes of the optical networking scene—brilliant, eccentric, mysterious, and rich as all get out," is how *Light Reading*, a fiber-optic industry journal, described him.[12] He wanted to focus on developing an all-optical networking device, but wasn't getting much support from the likes of Nettles and others on the board. It was only a matter of time before Huber and his company parted ways. In a huff, he quit six months after Ciena went public, and started a direct competitor, Corvis, which for the most part replicated Ciena's product offerings. Ciena sued Corvis on patent infringement charges later.[13]

In July 2000, Huber took his new company public. It raised $1.1 billion and, just days after its IPO, boasted a market cap of $38 billion, making it more valuable than General Motors. It was the biggest IPO ever by a start-up company that had yet to generate its first dollar of revenue! In fact, while its shares were being traded on the NASDAQ, Corvis' was yet to ship its first product. But at last Huber was able to be at the helm of his own operation.

Like other optical start-ups, Corvis gave shares to its customers, like

Broadwing, in exchange for getting orders from them. (Ironically, in early 2003 Corvis bought Broadwing for $250 million.) After the IPO, Huber started selling his stock almost immediately. In less than six months, he sold over $100 million worth of his holdings. This was, of course, before the company's value started to plummet. "I think the Corvis IPO was when things began to seem strange. It was getting kooky, because the company had a huge market capitalization, and there were questions about its customers. It was outrageous," said Kraus of US Venture Partners.

Like Kraus, another man who was feeling uneasy was Vinod Khosla. By the end of September 2000, it was easy to sense his growing unease. "The market is being reckless right now, and what you see going on right now is greed, and I know it sounds harsh. You will see a similar situation in optical space as in the dot-com space, and whenever there is a herd-like mentality it is not good. After all, if you don't understand the space, then you basically do not understand the difference between Lucent and Juniper," he told me. "You have a situation where a lot of people [are] investing in optical companies without understanding the basic trends or the technology."[14]

The reckless funding of optical companies could not have come at a worse time. With prices of bandwidth declining somewhere between 50 and 80 percent, some service providers were finding that making money was tough. In six months, the rest of the world would catch up with Khosla—two years too late, as always.

Optical Party Rocks On

But to the eternally optimistic, the Corvis IPO was a sign that all was well with the world of optics and communications. Even as Khosla warned the world about the bubble, the funding binge continued unabated. In the first week of October 2000, five start-ups raised nearly four hundred million dollars: Agility Communications ($70 million), CyOptics ($57 million), Lantern Communications ($59 million), Yipes Communications ($139 million), and Chiaro Networks ($101 million).

"Everything is getting funding, and I mean everything. Start-ups are all funded, even though they don't have business models, are late to the market, and lack seasoned management teams," David Aronoff, a general partner at the venture capital firm Greylock, told *Red Herring* magazine.[15]

That last surge in funding was like the gasps from a marathon runner coming to the end of his physical limits. A month later, in November 2000, Lucent dropped its bomb; it announced slowing sales of its optical equipment and decided to fire its chief executive, Rich McGinn. Meanwhile, Nortel announced a similar sales hiccup, and ICG Communications filed for bankruptcy. By early 2001, it would become clear that the market was gasping for air.

Paul Sagawa, an analyst with Bernstein Investment Research and Management, a division of the money management firm Alliance Capital, had released a report predicting that the new long-distance backbone networks would spend about $29.2 billion on equipment by the end of 2000, ten times the expenditure in 1995. Sagawa also said the CLECs would cut spending on new equipment from $8.1 billion in 2000 to $6.2 billion. Susan Kalla, another analyst, felt that the downturn in the carrier markets was not reflected in the equipment stocks and lowered her ratings on some of the top equipment and service providers. "The downturn in the carrier market is not yet reflected in the stocks of equipment makers," she told *Red Herring* in late 2000,[16] and predicted a shakeout in 2001.

Cisco, however, was in its own orbit. Unlike Lucent and Nortel, Cisco proved to be almost bulletproof. Having met or exceeded its revenue and earnings targets for almost 25 quarters, it seemed there was nothing Cisco could do wrong. The Cerent product line was proving to be a hit, and all looked good in San Jose even while competitors like Lucent and Nortel were in trouble because the end demand from service providers was melting faster than ice cream on a hot summer day.

However, between 1999 and 2001, Cisco's culture underwent a sea change. The executives started to get arrogant. In March 2000, the company had zoomed past General Electric and Microsoft as the largest company on the stock market. It was valued at *$578.5 billion* dollars. It was buying companies almost every two weeks.

Since it was sitting on billions of dollars in 1999, Cisco was buying its way into the market through vendor financing. It spent almost $2.4 billion on vendor financing. By constantly labeling them "old world" phone companies, Cisco managed to alienate many of the older, deep-pocketed telecoms, who decided to stick to Lucent or Nortel, or give upstart Juniper a chance. While the going was good, CLECs were still the stock market darlings, and Cisco did not feel the pinch of missing out on old-phone-company dollars.

Inside Cisco, there was a strange complacency. The company began hiring a lot of ex-IBM, Lucent, and Nortel employees, who were in it for the money. Many blame this influx of outsiders for the end of Cisco's entrepreneurial culture. "At one point in the meetings we would go in and there were new people every day, and you would not know them," recalled Junaid Islam, a former Cisco executive who worked closely in the acquisition team. "Arrogance and overconfidence took over, and nobody was paying any attention to customers who were complaining about the products not working."

Cisco insiders were suffering from a malaise called VIP—vesting in peace—and had no intention of working hard. Sure, Microsoft had its billionaires, but Cisco was more egalitarian—it had hundreds of centi-millionaires. They talked about private jet leases and about driving Ferraris instead of BMWs. The company was so focused on the stock price that almost everyone had a stock ticker displayed on their desktop computer. "It was so bad that Cisco had to put a ticker on the company intranet," recalled Islam.

New carriers, who were desperate for equipment, would buy Cisco gear even if it did not meet their requirements. They were under pressure from Wall Street to grow their networks. Otherwise they could not raise any more money from Wall Street. And when they started to fail—mostly because of too much debt and no demand—Cisco went into free-fall.

While the business was slowing down, Chambers insisted that the company could grow 30 to 50 percent per annum going forward. That was a staggering statement, given that Cisco had already grown to be a $19 billion (in annual sales) operation. The company's sales grew, year over year, 43 percent in 1999 and 56 percent in 2000, but only 18 percent in 2001. By 2002, sales were down 15 percent. Even assuming 30 percent growth, the company was betting that its sales could be $110 billion by 2005. Only one company in American history—General Electric—has done more than $100 billion in sales in one year, and it took GE 107 years, *Fortune* magazine has noted. Cisco wanted to sprint past the $100 billion mark in 21 years.

Unfortunately, many believed in Cisco's management—after all, this was a company that had become one of the first to garner a half-trillion dollars in market capitalization. To some extent, Cisco had suffered less than its rivals. Lucent and Nortel skidded when spending by the long-haul network operators and competitive local exchange carriers plummeted, but Cisco still had the stock to acquire companies, grow its business, and expand in new markets. About his company's acquisition strategy, Chambers told *Fortune*, "We have it down to a science. We could do 10 in a month if we needed to."[17] The *New Yorker* wrote: "Every bull market has its darling. In the nineteen twenties it was RCA. In the nineteen sixties it was Xerox. And in the nineteen nineties, it was Cisco."[18]

But on April 16, 2001, reality caught up with Cisco. The company announced that sales would be down for the quarter by 30 percent and that it would lay off 8,500 employees. It would write down $2.5 billion in inventory, which means that towards the end, Cisco's own management had no idea about demand. Chambers, like others in Cisco, believed that the company was safe from the troubles plaguing its rivals.

But the reality of failing CLECs and optical network operators had finally brought down the last standing titan of the broadband universe. From April 16, 2001, Cisco has seen its sales peter down to $18.9 billion at the end of fiscal 2002. At the time of this writing, its stock is trading for $14.89 a share. Cisco has refocused its energies on selling equipment to its time-tested customers—corporations, the government, and institutions. Thanks to this core business, Cisco is in a better position than Lucent and Nortel, and will likely be the king of equipment-makers in the decades to come. Meanwhile, it is still grappling with a glut of equipment on the market after the dot-com bust, which has created a huge gray market for Cisco gear.

Carl Russo, Don Listwin, and Kevin Kennedy have all quit Cisco. Raj Singh struggled with his next start-up, Roshnee, and is now financing movies, while Ajaib Bhadare has gone fishing—he can afford it. The indefatigable Khosla is still out there looking for the next big thing. And when he finds it, he will again be ahead of the game, by two years.

13

THE SWAMI OF THE BROADBAND BOOM

For almost five years in the 1990s, when it came to broadband, George Gilder was god. He preached the broadband gospel with the religious fervor of Pat Robertson, casting a spell on thousands of big and small investors, entrepreneurs, company chieftains, venture capitalists, investment bankers, and fund managers. His newsletter, *Gilder Technology Report*, prompted investors to forget the commandments of Benjamin Graham, the father of stock investment guidelines and valuation principles. His best-selling book, *Telecosm: How Infinite Bandwidth Will Revolutionize Our World*,[1] prognosticated a brave new world awash in bandwidth. Its thesis was that unlimited bandwidth would usher in an era of unlimited prosperity. Today, Gilder is the symbol of the telecom bust of the '90s. Once a celebrated author who was worth $1,000 dollars a minute, he is now dismissed as a mere stock tout and a techno-evangelist with a dead religion.

Gilder's mystical vision for a broadband future might have made for good reading (and, for nearly 65,000 subscribers to his newsletter, good investing) but it never really scratched beneath the surface to unveil the economic realities of the industry. Some of Gilder's favorite telecom companies—Global Crossing, Exodus Communications, and 360networks—are bankrupt, and his favorite medium of disseminating information to the masses, the *Forbes* magazine supplement *Forbes ASAP*, has fallen victim to an advertising downturn. He has lost his multimillion-dollar empire and former adoring fans. Like the companies he once passionately championed, Gilder is now staring down the financial abyss.

■ ■ ■

George Gilder was born on November 29, 1939, in New York City, to Richard and Ann Gilder. They later moved to the family farm in Massachusetts' Berkshire Mountains, where Gilder spent much of his childhood. Gilder comes from a well-known and once wealthy family; one of his great grandfathers was Louis Comfort Tiffany, the world-famous glassmaker. Another great grandfather was a friend of President Theodore Roosevelt.

Gilder's father, Richard, was killed in the Second World War, but David Rockefeller, a close friend and also his father's college roommate, took care of George, helping him secure admission at top schools like Exeter Academy and Harvard University. George was expelled from Harvard for being a sub-par student, but was later readmitted and graduated with a degree in government studies. While in his 20s, Gilder lived in New York's East Village and led the life of a bohemian writer on the verge of insolvency. While in his 30s, Gilder wrote two controversial books, *Sexual Suicide* and *Naked Nomads*, which made him the man feminists loved to hate. *Time* magazine and the National Organization of Women named him "Male Chauvinist Pig of the Year."

Gilder lived on the fringes until 1981, when he published *Wealth and Poverty*, the book that changed his life. It was an ode to entrepreneurialism and its ability to cure societal ills like poverty. The book found a ready audience in two camps—the supply-side economics–driven Reagan Administration and Silicon Valley's entrepreneurs. It sold more than a million copies and made Gilder a supply-side guru. It is said that he was one of the authors whom President Ronald Reagan quoted most often. This was the first step towards Gilder's ultimate station in life: the Prophet of Boom. (It also earned him the ire of author and philosopher Ayn Rand, who devoted her last public speech to denouncing George Gilder. Gilder's libertarian views perhaps made her mad!)

In *Wealth and Poverty*, Gilder devoted a full chapter to Silicon Valley pioneers. Gilder had become intrigued by semiconductors and decided to study physics at the California Institute of Technology under the tutorage of Carver Mead, a well-known physicist and chip guru who also founded 25 companies, according to *Red Herring* magazine. Gilder managed to learn the intricacies of semiconductors and in 1989 penned his first pure technology book, *Microcosm*.

The book, which focused on Intel and on Moore's Law—which says that the power of chips will double every 18 months—made Gilder an overnight sensation among the libertarians in Silicon Valley, where almost everyone read him. The book became an unlikely bestseller. A year later, Gilder published another book, *Life after Television*, which predicted the demise of television and the rise of the personal computer. He talked about how a world of glass and light, and chips, would make television redundant. Many of those early predictions have come true. No, television hasn't died, but it is changing. It was sometime after the publication of *Life after Television* that Gilder met Will Hicks, a preeminent scientist who specialized in fiber-optic technologies.

By 1993, Gilder was a convert to the new religion of fiber optics and was ready to spread the gospel. He choose *Wired* magazine, where he was an occasional contributor, as his medium to start the optical revolution. "We're going to gain access to the 25,000 gigahertz of capacity that's in each of the three windows in infrared spectrum that work with fiber optics. With 25,000 gigahertz, you get the equivalent of the number of phone calls in America during the peak moment on Mother's Day," he told *Wired's* founder and editor, Kevin Kelly.[2] "My thesis is that bandwidth is going to be virtually free in the next era in the same way that transistors are in this era. It doesn't mean there won't be expensive technologies associated with the exploitation of bandwidth."

Gilder predicted there would be hundreds of small companies that would get started and make his vision a reality. His predictions would come true and help propel Gilder to the forefront of the broadband revolution of the 1990s. "What makes Mr. Gilder more than your garden-variety technology pundit (and therefore more dangerous) is that he knows what he is talking about. His stuff is dense. It's full of technical arcana. It is learned, purposely literate, and it oozes legitimacy," Paul Kedrosky, a professor of business at the University of British Columbia, later wrote in an editorial for the *National Post*, a Canadian business newspaper.[3]

■ ■ ■

By 1995, Gilder still a writer, was beginning to make some serious money. People were beginning to pay him thousands of dollars for speaking engagements, and the royalty payments for his books were growing. He needed professional help to keep his new financial house in order, and Chuck Frank and David Minor, two money managers, were perfect for the job. Later Frank and Minor suggested that the three of them go into business together as the Gilder Technology Group.

Their original idea was to publish research for Wall Street brokerages, but Wall Street firms showed no interest, so that idea went nowhere. Gilder, who at the time was writing for *Forbes* and *Forbes ASAP*, came up with an idea to write a monthly technology newsletter. He pitched the idea to Steve Forbes, and a deal was struck: Gilder would write the newsletter and Forbes would publish, market, and distribute it. The two companies, Forbes and Gilder Technology Group, would divvy up the revenues.

At the time, the Telecommunications Act of 1996 was being formalized and would go into effect shortly before the launch of the newsletter, named the *Gilder Technology Report*. The timing was perfect. There was no literature available on the technologies that would boom because of the telecom deregulation. Fiber optics was still a mostly unrealized science, and there were no experts who knew broadband as well as Gilder. And if that wasn't enough, he was also advising the likes of House Speaker Newt Gingrich (R-Georgia) on broadband issues. Like the term *microcosm*, which Gilder derived from the word microprocessor, he came up with a new term, *telecosm*, from telecommunications.

"The law of the microcosm is now potentially converging with the law of the telecosm. This law ordains that the value and performance of a network rise apace with the square of the increase in the number and power of computers linked on it. As these forces fuse, the world of computers and communications can ride an exponential rocket. The 50 million new computers sold into America's homes and offices over the last two years guarantee a huge market for broadband networks," Gilder wrote in *Forbes ASAP* in April 1995.[4] Such pronouncements were very well received and were featured prominently in the newsletter, which was launched in mid-1996 with an initial print run of 8,000. It was a hit.

It is not difficult to see the parallels between what Gilder's newsletter preached and the business plans of early stage companies like Qwest and Global Crossing. In the fall of 1997, Gilder held the first annual Telecosm Conference at the Ritz Carlton Hotel in Palm Springs, California, and nearly 350 people paid $4,000 to attend. By then, the newsletter had a circulation of 10,000, and Gilder was an in-demand speaker who earned $50,000 per speech and became fabulously wealthy.

How Optics Was Sold

By 1997, it was becoming obvious that, while Gilder could paint a compelling picture with his description of optical and broadband technologies, it was difficult to take that technobabble, plug it into spreadsheets, and build business models with it. Everyone needed forecasts and hard numbers that broke down industry sectors into dollars and cents. If a Wall Street firm needed to sell an initial public offering or build stock price models, they needed some sort of market forecasts. Similarly, venture capitalists and entrepreneurs needed hard numbers to get backing for start-ups.

Enter Ryan Hankin and Kent Inc. (RHK), a tiny, south San Francisco–based research firm founded in 1991 by three former telecom executives. For years, the firm had existed on the fringes, but its staff had accumulated in-depth knowledge of the optical and broadband business. This firm was perhaps the only organization that had a handle on the dollars-and-cents aspect of the optical business.

RHK's big moment came in 1997, when it predicted that total demand for optical equipment in that year was going to be around $1.5 billion; at the time, most other pundits were predicting demand would be around $300 million. At the end of 1997, when the total tally came in at $1.45 billion, RHK's star rose among Wall Street firms and venture capitalists.

Research firms like RHK play a vital role in the selling of technology. They translate the technical mumbo jumbo into plain English; they legitimize the products and technology with bullish and often overoptimistic predictions of the total market size. In short, they make everyone see the world through their rose-tinted glasses. I remember that from 1997 on, most company officials who visited the Forbes.com offices in New York's trendy Silicon Alley used RHK charts and data in their presentations. The firm's analysts were often quoted in news reports, and they spoke with the kind of authority that mere mortals—including lowly journalists—lacked. Journalists would call RHK and get easy answers to questions about the size of various telecom markets. By devoting inches of newsprint to their predictions, the media turned analysts like RHK and Gilder into telecom authorities. The personal computer industry depends on raw data from research firms like International Data Corporation, Gartner Group, and Dataquest. Broadband and optics needed its own research group, and RHK fit the bill perfectly.

RHK came out with forecasts that were too optimistic, while other analysts, like Dr. John McQuillan (of NGN Ventures), sat on the technical advisory boards of the firms, thus compromising their integrity. Some accepted stock in start-ups that they were supposed to analyze. Unlike Wall Street firms, where the issue of separating investment banks' research arms from their banking divisions has caught the attention of New York Attorney General Eliot Spitzer, there are no rules for these so-called market research firms.

A former employee of RHK told me that his former employer accepted stock from companies that included Juniper Networks. The firm, which saw its revenues increase almost 125 percent from 1999 to 2000, accepted venture backing from Crescendo Ventures, which was one of the top investors in optical start-ups. Former RHK employees have mentioned that one of the three founding partners, Peter Hankin, started a venture fund called the Infrastructure Fund with help from another Silicon Valley venture firm, Interwest Partners. "The way the carry works to the best of my knowledge is that part of it goes to Peter and part of it goes to RHK or to its analysts," a former RHK employee told me.

Tom Noelle, a veteran telecom analyst and principal of CIMI Corporation, a Voorhees, New Jersey–based research firm, said that during the boom years, he would often be approached by companies to join their technology advisory boards and help them sell their vision of the product. "In order to get any money I would have to sell that story," said Noelle. With more than 2,000 start-ups, analysts would field these kinds of calls from companies every day. They became the filter for the popular media and customers. If an industry analyst decided that he liked a company (or if he was well compensated in the form of shares) he would talk up that company to reporters—and to potential customers.

"I think if you go back in history, there was this technology [called] ATM [Asynchronous Transfer Mode], which was entirely sponsored by [the] media and the analyst community," says Noelle. The hype around ATM technology, which was an expensive way to send data and voice traffic over the networks, was so high that it helped sell the products and also helped push the stocks of ATM companies. "What that did [was] it showed Wall Street and [the] venture capital community that it was easy to flip a company even if there was no success or even [a] semblance of success," Noelle added.

As a reporter at the time at Forbes.com, it took me some time to figure out that the game was rigged. Since the technologies were so hard to understand, poor chemistry majors like me had to rely on the "experts." As a result, these independent industry analysts became demigods for the start-ups. Given that most start-ups were vying for attention and mind-share, offering shares to analysts was a small price to pay. "I was regularly offered founder shares. It was done to make the key figures as insiders, and that was wrong. It was fraud and complete conflict of interest. I think no one was actively looking at conflicts of interest," said Noelle, who always declined such offers. "The greed was ubiquitous, otherwise key figures could not have gotten away with it," remarked Lawrence Gasman, president of Communications Industry Researchers, another analyst firm. He was pointing out that everyone, including the media, the analysts, and George Gilder, overlooked the ethical missteps of the 1990s.

The Gilder Effect

With nearly 65,000 subscribers for the *Gilder Technology Report*, Gilder had immense influence over the future of the companies he chose to write (or not write) about. From 1999 to 2001, people waited for their copy of the newsletter just as grown men wait for their copy of *Maxim* magazine. To them, the publication was a one-way ticket to riches. Most of Gilder's subscribers would skip the highly technical mumbo jumbo and skip right to the back of the newsletter, where Gilder included a list of about 30 companies he called the "Telecosm Technologies." If a quick scan revealed any additions or deletions, the next stop for Gilder's flock of 65,000 was their online brokerage account.

And they were well rewarded for their efforts. In 1999, there was no pundit like Gilder. Six of the top performing stocks in the Standard & Poor's Index (including Broadcom, Qualcomm, and Sun Microsystems) were Gilder's Telecosm picks. Qualcomm alone was up 2,618 percent, while others managed to rise more than 200 percent for the year. In December 1999, when Gilder added Novell to the list, the beleaguered company saw its market valuation jump by $2 billion. Another stock that did quite well was JDS Uniphase, up 830 percent. "I had six of the top nine stocks on the S&P, and four of the top eight on the NASDAQ," Gilder later told *Wired*.[5]

Gilder's reputation was further bolstered by his unrelenting and unflinching support for a new technology called DWDM, or Dense Wavelength Division Multiplexing. This technology can help slice a ray of light into many different colors (or wavelengths), and then carry different information on each color or wavelength. Gilder championed a little-known company called Ciena, an early leader in the DWDM business. "Its February IPO was the most important since Netscape (market cap at the end of the first trading day: $3.4 billion). Why? Ciena is the industry leader in open standard WDM gear," wrote Gilder in *Forbes ASAP*.[6]

Gilder saw Ciena as one of the key companies that would realize his vision of an all-optical network. "The all-optical network will triumph for the same reason that the integrated circuit triumphed: It is incomparably cheaper than the competition. Just as the electron rules in computers, the photon will rule the waves of communication," he predicted.[7]

In April 2000, when most other technology stocks were nose-diving, George waxed eloquent about a new company: Avanex, a Fremont, California–based fiber-optic component company. As a result, the company's stock soared. Avanex was trading at $55 a share when Gilder blessed it, and a week later, it was going for $140 a share.

In May 2000, when Gilder added Terayon Communications Systems, a cable modem maker, and Agilent Technologies, a high-tech equipment maker spun out of Hewlett-Packard, to the anointed list, their stocks jumped 17 percent. In October 2000, he blessed Exodus Communications, a Web hosting company, and its stock jumped 13 percent. The company has since gone under.

When Gilder soured on a company, like he did on the network backbone provider Level 3 in November 2000, the stock plummeted in nanoseconds. On November 13, 2000, Level 3 shares fell $5.81 a share to $30.25 a share. With the omission from his list, Gilder wiped out almost $2.5 billion from Level 3's market capitalization. "A man whose slightest utterance can move stocks," is how the *Wall Street Journal* described Gilder in a very flattering profile. Wall Street professionals called his awesome power to move markets "The Gilder Effect."

Even in his best-selling book, *Telecosm: How Infinite Bandwidth Will Revolutionize Our World*, Gilder included a list of companies he called "Nine Stars of Telecosm." These included names like JDS Uniphase and Qualcomm. "What Gilder really seems to tactically admit in *Telecosm* is that his ultimate role in American Discourse is basically as a stock-picker, and he's OK with that," wrote Rob Walker in his review of the book for the online magazine Slate.[8] Gilder actually admits that on his web site, where the first thing you see is: "Can you turn $10,000 into $17,708,483?"

At the peak of the stock market boom, in the days before the dot-com bubble burst, Gilder decided to buy out his partners for about $8.5 million and renamed his company Gilder Publishing. He had started other newsletters focused on different sectors, like storage and digital power components. He also got ambitious and bought the *American Spectator*, a conservative magazine, for $2.5 million. Gilder later said that he spent a total of $11 million at a time when he could least afford it, but he was hopeful that his company would raise money from the public markets. Hambrecht & Quist and Merrill Lynch were vying to take Gilder Publishing public.

"There was talk of a $200 million valuation. I thought we were rich," he later told *Wired*.[9] And they were—the company was bringing in around $20 million a year. In an interview, Gilder admitted that he was worried about the valuations of some of the companies he had lavished praise on and wanted to warn his newsletter subscribers. He did not do it because he was worried that he could precipitate a sell-off that was not really there. The effect that had made Gilder a superman was also proving to be his kryptonite.

Just Don't Sell Us

No company benefited from Gilder's endorsements more than JDS Uniphase, a maker of fiber-optic components that are the very basis of all new networking equipment—sort of like vodka in a martini. JDS was precisely the kind of company Gilder liked to champion. It had a quirky CEO, Jozef Straus; it made cool-sounding products (erbium-doped amplifiers); and it was a virtual unknown in the world of business. So when Gilder called it the "the Intel of the Telecosm," voilà, its rocket rise to the top began.

Jozef Straus is a bohemian iconoclast: He is known to kiss reporters at annual meetings, and he always sports a beret. Straus is the founder of the company that started its life in 1981 as JDS-Fitel. Straus was born in July 1946, in a tiny farming village in what used to be Czechoslovakia, to Jewish Holocaust survivors. He was studying physics and nuclear science in Prague in 1968, when he left the country three weeks before the Soviet invasion. He ended up in Canada, where he attended the University of Alberta in Edmonton and completed his Ph.D. in physics in 1974. He then joined Bell Northern Research's optical research group in Ottawa, where he got all the skills that led to the next step of his life: JDS-Fitel.

He co-founded JDS-Fitel in 1981 and joined the optical component company full-time five years later. For the next decade or so, JDS lived on the fringes. Then the optical boom happened, and the company saw its sales go up sharply. In 1999, Uniphase, a San Jose, California–based rival, came calling.

Uniphase was started in 1979 and was earning a modest living by making lasers for chip makers, supermarket scanners, and other low-end devices. In 1992, the company hired Kevin Kalkhoven as its chief executive. The tough-talking Aussie decided to make a bet on telecom-related optical products and changed the direction of the company. It went out and bought small optical component companies and completed a makeover that suddenly made it a decent-sized player in telecom markets. "What we did was move aggressively [to acquire companies] before other people caught on to the idea. And we did it while things were cheap," he later told *Fortune*.[10]

In 1999, JDS-Fitel and Uniphase announced a $5 billion merger—which Straus celebrated by giving the 20,000 employees of the new combined company his trademark berets. At the time, Gilder had liked both companies and had mentioned their products often in his newsletter. Together, he thought JDS Uniphase could do no wrong. In late 1999, the stock of the combined companies, which traded under the ticker JDSU on the NASDAQ, caught fire. By early 1999, it was trading at $10 a share and it hit an all-time high of $146.53 a share (split-adjusted—before the split it was $293.06) on March 6, 2000.

The company went on a buy-and-grow-big fast track, buying 17 companies for more than $60 billion after beginning life as a new company. It was the WorldCom of fiber optics. It bought rival E-Tek Dynamics for $15 billion in June 2000. But right in the middle of the merger, in May 2000, Kalkhoven decided to hang up his Oxfords and strolled into the sunset with a cool $250 million in his wallet! He could now go back to his first love—racing cars. In an interview with the Australian newspaper the *Bulletin*, he quipped, "I made a lot [of money]. I've just bought the latest Gulfstream IV [business jet] and it's absolutely gorgeous."[11]

But Kalkhoven's exit did not stop the JDS monster—the company's executives decided to buy out competitor SDL Inc. for a whopping $41 billion in July 2000. JDS was paying a 50 percent premium over SDL's then stock price of around $30 a share. JDS could afford it—its stock at the time was around $128 a share, thanks to a little help from friends like Gilder. At the time, JDS Uniphase had a market capitalization of $181.32 billion, even though its sales were $1.4 billion and it had posted a loss of $904 million for the 2000 fiscal year.

Of course, JDS Uniphase overpaid for most of its acquisitions. For instance, it bought a company called SIFAM in December 1999 for $90 million, a figure that was twice the total market for its fused-fiber optical components. Needless to say, JDSU's $41 billion acquisition of SDL far exceeded the latter's addressable market at the time of the announcement. (See Table 13.1.)

To get the JDS/SDL merger deal approved by the U.S. Department of Justice, the company had to sell off some operations. Nortel bought the Zurich facility and an operation in Poughkeepsie, New York, for $2.5

Table 13.1 JDS Uniphase Corporation—Select Acquisitions
Buy and grow today, worry later.

Announced Date	Target	Size ($mm)
March 2000	Cronos Integrated Microsystems, Inc.	$ 750.0
June 2000	E-Tek Dynamics, Inc.	19,000.0
July 2000	SDL, Inc.	41,143.6
December 2001	Int'l Business Machines, Optical	340.0
April 2002	Scion Photonics	43.0
	Total	$61,276.6

Source: Capital IQ

billion. But as far as the Zurich operation was concerned, JDS was smart to get rid of it. An analyst who had toured the facility right after the JDS Uniphase merger quipped, "The facility was built as a laboratory, not as a production environment."

Here was something even Gilder didn't catch. Swiss law dictated that the Zurich operation had to have natural light in every room in the building. But natural light causes problems in the processing of wafers, and therefore must be filtered. As a result, this research lab with windows everywhere was hardly a first-rate production environment. "When JDSU had to sell it to Nortel, I suspect some were secretly happy," this analyst said.

The stock party continued through much of 2000 and 2001. Still wondering how long this boom would last, the 56-year-old Straus decided to take about $150 million off the table by exercising his JDS stock options. He had often remarked that "the emperor has no clothes," and knew that the boom would not last forever. He knew that the whole bull market was unreal, and perhaps that's what prompted him to sell his shares when the going was good. Aside from Straus, the company compensated other executives quite well—Don Scifres, the JDS co-chairman and SDL's chief executive before the merger, got $75 million in cash, while chief operating officer Gregory Dougherty (also from SDL) received $75 million as a signing bonus.

In April 2001, reality caught up with JDS Uniphase. Being at the bottom of the optical food chain, it was feeling the effects of the slowdown, the last. The company warned investors about its fiscal future and in the second quarter of 2001 reported sales of $925 million and profits of

$208 million. It has been pretty much downhill since then. The company, which had sales of over $1.4 billion in 2000, is now facing shrinking revenues. It has taken $65 billion in write-offs. The stock is down to a mere $3.13 a share.

What Gilder really did in the case of companies like Avanex, JDS Uniphase, Broadcom, and several others, was create a mirage. These were fantastic growth plays, which in turn allowed insiders to sell out their shares at high prices. I don't think he did it intentionally, but Gilder really did not say that things were getting kooky. He should have mentioned that JDS Uniphase was overvalued and that perhaps investors should take some of their money off the table. "In retrospect, it's obvious that I should've subtly said, 'Hey, things have gotten out of hand at JDS Uniphase, and it's not worth what you'd have to pay for it,'" he later told *Wired*.[12] In other words, if the stock goes down, he doesn't lose anything—but when it goes up, Straus can buy another beret and have it decorated with diamonds and pearls.

Dis-connected

JDS Uniphase was suffering from the same problems as the rest of the industry. The slow decline of Gilder's telecosm began with the Nortel announcement in February 2001. That brought the crisis home for Gilder. The American Stock Exchange's Networking Index, which is formed of some of the top telecom and broadband names (many of them Gilder favorites), was down almost 32 percent in the first two months of 2001. In addition, stocks like Corvis, Avanex, JDS Uniphase, and Global Crossing were being pounded in the stock market.

But still George Gilder did not give up his favorite telecosm companies. In his February 2001 column for *Forbes ASAP* he wrote, "The current disdain of the press and Wall Street wise guys gives you the chance right now to buy a stake in the stratospheric future of communications at the price of a pedestrian blue chip. In a sense, this is a double-or-nothing endorsement of Global Crossing, which remains at least two years ahead of 360networks. Today, there is no economy but the global economy, no Internet but the global Internet, and no network but the global network. Global Crossing and 360networks will battle for worldwide supremacy, but in a trillion-dollar market, there will be no loser. Despite the foibles of its management and its revolving CEOs, Global Crossing remains the world's best-situated telecom company."[13] Global Crossing was bankrupt less than a year later.

Gilder loved Global Crossing so much that he, too, had invested in the company. In 2001, he was fighting a losing battle. His favorite companies were seeing their revenues drop sharply, and there was bankruptcy after bankruptcy. In January 2002, when Global Crossing filed for bankruptcy, Gilder was financially ruined. Today, there is a lien against his house.[14]

"There were so many moving pieces in the middle of the storm, and I couldn't figure it out," Gilder said in an interview in November 2002. "In 1997 I started to warn against the [global] deflation and that telecom debt was going to be a problem." But he admits he did not say it enough. "All my carrier choices went bankrupt. The carrier model, for whatever reason, failed almost entirely," he added. He blames deflation, high interest costs, and excessive telecom regulation in the United States for the broadband bust. He still defends the companies he wrote about, offering the argument that "I don't think these were criminal conspiracies."

It is remarkable how Gilder's brief life of fame and fortune followed the trajectory of the stocks and of the companies he devoted his life to. His star rose with the stock market, and fell with it. But still it is hard to find anyone who has sympathy for Gilder—not even those who read his newsletters. Call it a strange twist of fate: He spent too much money when he should not have; much like Excite@Home or Teligent, he grew too fast and his staff did not have a handle on the finances of Gilder Publishing, just like in WorldCom. "Bad behavior or misleading wasn't deliberate, because most were misleading themselves and they ended up misleading others. George Gilder is one of those—he was mistaken," said Lawrence Gasman of Communications Industry Researchers (CIR).

While Gilder was absolutely spot-on as far as the technology was concerned, he made three strategic mistakes. First, he made everyone believe that the utopian future he espoused was here and now, even though in reality the all-optical nirvana is still some years off. Second, he did not take into account the regulatory dimension to the whole broadband bubble. He believed, in his naiveté, that the Bells, the ultimate chokeholds on the local loop, would simply roll over and let competitors take over the market. Today he says he knew the regulatory problems but did not write about them enough. Why? Because it did not fit in with the boom-culture of the 1990s. Gilder's folly was that he believed that the technology could ultimately win over market realities.

And third, the biggest mistake of all was that he started making stock recommendations, never taking into account the financial dimension of the boom—the debt, the interest on loans, and the ultimate demand. He believed whatever the likes of Global Crossing's Gary Winnick, World-Com's Bernie Ebbers, and others told him. He is a futurist, not a reporter—he didn't dig deep enough, or perhaps he did not want to. When asked whether he would stop recommending stocks, Gilder simply said, "I have four newsletters, and they are in the stock picking business," pausing for a second and adding, "It is painful."

How much is Gilder to blame? After all, other pundits, like the research firm RHK, made irresponsible forecasts that made the market seem bigger than it really was. Even other so-called analysts, who were accepting stock from start-ups, or those who started to make venture capital investments, did the same. What about our collective greed and desire to look for the next hot stock? Five years from now, Gilder's predictions might actually come true. The Gilders of the world and the entrepreneurs represent the infinite possibilities of technology. Today, in the middle of a telecom depression, more than ever we need Gilder, but we *don't* need his stock picks.

EPILOGUE: THE END GAME

Somewhere in the middle of writing and researching this book, for a very brief while, my cable modem went on the fritz. The wait for the connection to come back online was as excruciating as the cravings of a smoker on a red-eye from San Francisco to New York. Being disconnected from the service made me realize that, despite all the doom and gloom surrounding the broadband business today, in five years an always-on high-speed Internet connection will be as much a part of our lives as electricity. Too bad @Home management wasn't smart enough to make the right decisions and be around to take my $50 a month.

Between 1996 and 2001, what went wrong was the execution of the broadband dream. The six-year-long broadband bubble was the result of a here-and-now culture where all of us became obsessed with our "net assets" and forgot about our "net worth." From the very start, the deregulation of the telecom business was flawed; the Telecommunications Act of 1996 was drafted without taking into account the coming Internet tsunami. Baby Bells, which are among the highest contributors to politicos, ensured that they have a home-field advantage.

Greedy chief executives undermined the American free market system by stuffing their coffers; Wall Street encouraged the management of broadband companies to adopt short-term metrics; and investors simply played along. And the little guys just got slaughtered. The greedy chief executives who failed upwards should be penalized, and I would not shed a single tear if any one of them is sent to the big house. They deserve it.

But something good came from the bubble as well. Baby Bells, who had been sitting on the DSL technology for almost a decade, were forced to roll out the service because Covad and NorthPoint put pressure on their T-1 business. Sure, the Bells won, but so did the consumers. @Home's early success prompted cable companies to make Internet access a priority. Long-distance phone prices are down into the single digits.

While these developments will not replace the lost dollars in the 401(k) of a telecom worker, they offer some hope for the population at large.

Despite the current crisis in the broadband business, I am a lot less despondent today than I was when starting work on this project. We should take our cue from far-off places like Japan, South Korea, and China, where millions get their television, phone, and Internet service over a high-speed DSL line.

If you are looking for a glimpse into this high-speed future, I suggest you visit Sioux Falls, South Dakota. It may be an unlikely place to look for the broadband society, but it is real. Sioux Falls is one of the half-dozen locations where Seattle-based Monet Mobile has deployed a high-speed wireless network that allows folks to stay connected to the Internet wirelessly at speeds that are as fast as DSL-based Internet access.

If Sioux Falls is too far, then visit Bryant Park in New York; order a latte and enjoy Web surfing for free. And now imagine a world where there are thousands of such networks, and it begins to dawn on us that the telecom bubble that burst was part of a painful but necessary cycle. Like its predecessors, the radio, railroad, airline, and automobile bubbles, the broadband bubble will soon become a distant memory.

Sure, the industry will suffer for a couple more years, but by then entrepreneurs—the very essence of the American capitalist system—will figure out a way to use that bandwidth. Steve Jobs of Apple Computer wants all of us to exchange digital photos and videos; that will consume some bandwidth. Some say that a new era of grid computing will dawn.

The telephone industry is undergoing a massive change, a change which many smart people compare to the shift from a minicomputer to a personal computer. As with that shift, there will be new companies and new businesses will emerge. Without millions in venture capital, start-ups today are bootstrapping their way up. Herman Miller chairs are out, and frugality is hip once again.

And that is a good start!

ACKNOWLEDGMENTS

We are all but a sum of many parts, and so is this book, which has been written with the help of many who wish to go unnamed; but there are a few who have offered insights, encouragement, and suggestions along the way.

Red Herring editors have been particularly supportive of this endeavor, especially Executive Editor Duff McDonald, who put up with me showing up bleary-eyed at noon. Blaise Zerega (now at *Wired*), Jason Pontin, and Anthony Perkins, editors at *Red Herring*, have always been enthusiastic supporters. My former editors at *Forbes* and Forbes.com—Mathew Schifrin and David Churbuck—showed faith in my reporting, particularly when I had none. They taught me to take off the rose-tinted glasses and be very cynical.

I am grateful for the support of Jeanne Glasser at John Wiley & Sons for taking a chance on an unknown like me; and to my agent, Joni Rendon, for believing that telecom is not such a boring story after all. Of course, my parents finally get a chance to figure out what I do for a living! And my many thanks to my very dear friends Shailaja Neelakantan and Joanna Pearlstein, who finally helped me make sense out of the madness. They are the sole reason I was able to deliver this manuscript on time.

Many helped in writing this manuscript; others were extremely generous with their time and insights: Brian Thompson; Phil Quigley; Danny Bottoms; Mark Langley; Bill Lesieur; Chris Noelle; Christine Heckart at Juniper Networks; Lawrence Gasman at CIR; Susan Kalla; Ravi Suria; Patricia Lee; Ravi Malik; Bob Olson; Aman Kapoor; Julie Meyer; Rakesh Mathur; Bundeep Singh; Jon Bayless and Amra Tareen at Sevin Rosen Funds; Steve Kraus at US Venture Partners; Sanjay Subedar at Storm Ventures; Steve Kaufmann at VCPR; and Charles Dubow and Michael Noer at Forbes.com. There are many others who wish to remain anonymous, and to them I am eternally grateful.

Data providers for this book are the most reliable services: Thomson

Financial, Venture Economics, Capital IQ, Baseline, Edgar Online, Lexis-Nexis, Pacer, Findlaw, and Capital Access.

Of course, this book was not possible without a high-speed Internet connection and Google, still the gold standard in Internet search. It is amazing what you can find on the Internet.

More than anything else, this project was not possible without the help of Deepak Talwar. He came up with the clever title *Broadbandits*. The rest is just a story.

O. M.

APPENDIX A:
CASH & CARRY

Carrier insiders sold out quickly; the more they sold, the more likely the company would go bankrupt.

Ticker	Company Name	Total Market Value
FON	Sprint Corp. Fon Group	$3,293,892,919
GBLXQ	Global Crossing Ltd.	2,625,114,819
NXTL	Nextel Communications Inc. New	1,213,052,116
PCS	Spring Corp. PCS Group	1,200,604,206
MCLD	McLeodUSA Inc.	1,089,012,641
WWCA	Western Wireless Corp.	666,040,320
XOXOQ	XO Communications Inc.	648,710,802
COVD	Covad Communications Group Inc.	602,465,519
GSTRF	Globalstar Telecommunications Ltd.	598,760,911
TPC	Triton PCS Holdings Inc.	549,911,933
WCOM	WorldCom Inc. New WorldCom Group	538,011,958
T	AT&T Corp.	515,881,442
TGNTQ	Teligent Inc.	385,061,509
LCCI	LCC International Inc.	236,957,884
TALK	Talk America Holdings Inc.	217,945,253
NPNQ	Northpoint Communications Group Inc.	184,229,519
VZ	Verizon Communications Inc.	176,379,647
SBC	SBC Communications Inc.	172,730,925
APS	Alamosa Holdings Inc.	165,881,964

Ticker	Company Name	Total Market Value
CYCL	Centennial Communications Corp. New	$ 152,528,554
RMHT	RMH Teleservices Inc.	140,572,922
RSLCF	RSL Communications Ltd.	129,669,429
PCSA	AirGate PCS Inc.	109,372,563
DCEL	Dobson Communications Corp.	89,776,894
MFNXQ	Metromedia Fiber Network Inc.	88,018,031
CTL	CenturyTel Inc.	75,700,123
ESPIQ	e.spire Communications Inc.	72,946,829
PGEXQ	Pacific Gateway Exchange Inc.	69,644,542
AT	ALLTEL Corp.	68,242,303
CTCI	CT Communications Inc.	64,285,448
CZN	Citizens Communcations Co.	63,647,050
BLS	BellSouth Corp.	59,073,758
RTHMQ	Rhythms NetConnections Inc.	51,110,833
GNCMA	General Communication Inc.	51,027,983
ACTL	Actel Corp.	49,239,413
BRW	Broadwing Inc.	46,559,654
WLNKQ	WebLink Wireless Inc.	41,844,264
WCIIQ	Winstar Communications Inc.	39,596,164
MCLLQ	Metrocall Inc.	38,821,525
DAVL	Davel Communications Inc.	36,927,429
CVST	Covista Communications Inc.	31,016,053
LTBG	Lightbridge Inc.	30,899,910
TDS	Telephone & Data Systems Inc. DE	23,717,041
USM	United States Cellular Corp.	23,648,547
FCOM	Focal Communications Corp.	23,486,789
ITCD	ITC DeltaCom Inc.	23,441,481
GOAM	GoAmerica Inc.	19,139,535
MTON	Metro One Telecommunications Inc.	16,602,724
CLEC	US LEC Corp.	16,426,646
UNWR	US Unwired Inc.	13,982,663
CTCO	Commonwealth Telephone Enterprises Inc.	12,330,141
PRTL	Primus Telecommunications Group Inc.	11,201,010

Ticker	Company Name	Total Market Value
NPLSQ	Network Plus Corp.	$ 10,601,472
MPWRQ	Mpower Holding Corp.	8,455,484
SURW	SureWest Communications	7,304,525
STGC	Startec Global Communications Corp. DE	5,497,314
ADELQ	Adelphia Communications Corp.	4,373,040
ABIZQ	Adelphia Business Solutions Inc.	3,847,800
USCM	USCI Inc.	2,909,091
NTLO	NTELOS Inc.	2,088,164
RCCC	Rural Cellular Corp.	1,912,886

Source: Thomson Financial

APPENDIX B: BANK BALANCE BUILDOUT

Broadband buildout also built the bank balances of many equipment company executives.

	Ticker	Executive	Sales
		Total Sales, 1997–2001*	
Foundry Networks	FDRY	Bobby R. Johnson Jr.	$411,176,466
JDS Uniphase	JDSU	Jozef Straus	181,263,318
Brocade/Avanex/	BRCD,AVNX,		
Glenayre	GEMS	Gregory L. Reyes	383,413,331
Brocade	BRCD	Seth D. Neiman	319,422,440
Powerwave			
Technology	PWAV	Alfonso G. Cordero	305,830,767
Cisco Systems	CSCO	John T. Chambers	296,189,993
JDS Uniphase	JDSU	Kevin N. Kalkhoven	251,783,050
JDS Uniphase	JDSU	Danny E. Pettit	205,451,533
Enterasys	ETS	S. Robert Levine	147,436,540
Cisco Systems	CSCO	Gary J. Daichendt	137,690,846
Scientific Atlanta	SFA	James F. McDonald	135,967,119
Brocade	BRCD	Kumar Malavalli	131,483,840
3Com	COMS	Casey G. Cowell	125,059,313
Cisco Systems	CSCO	Larry R. Carter	117,981,700

*Does not include transactions reported as Indirect in nature on the Forms 4.

(Continued)

Total Sales, 1997–2001*

	Ticker	Executive	Sales
Sycamore Networks	SCMR	Gururaj Deshpande	$117,237,400
Cisco Systems	CSCO	Carl Redfield	111,058,171
Juniper Networks	JNPR	Pradeep Sindhu	108,618,440
JDS Uniphase	JDSU	Joseph Ip	103,995,124
Finisar	FNSR	Gregory H. Olsen	102,694,760
JDS Uniphase	JDSU	M. Zita Cobb	100,065,582
3Com	COMS	Ross Manire	99,448,278
3Com	COMS	Janice M. Roberts	97,166,981
Sycamore Networks	SCMR	Richard Allan Barry	96,142,450
Cisco Systems	CSCO	Edward R. Kozel	93,843,155
Sycamore Networks	SCMR	Chikong Shue	90,997,953
Cisco Systems	CSCO	Judith Lenore Estrin	90,762,249
Foundry Networks	FDRY	Lee Chen	87,126,367
Juniper Networks	JNPR	Peter Wexler	86,905,690
Juniper Networks	JNPR	Steven R. Haley	83,490,096
Ciena	CIEN	Steve W. Chaddick	73,105,094
3Com	COMS	John McCartney	77,371,641
Sycamore Networks	SCMR	Daniel E. Smith	76,365,600
Enterasys	ETS	Craig R. Benson	75,303,320
Sycamore Networks	SCMR	Eric Swanson	66,979,328
Brocade	BRCD	Michael J. Byrd	65,809,107
3Com	COMS	Michael S. Seedman	63,615,580
JDS Uniphase	JDSU	Anthony R. Muller	61,948,710

*Does not include transactions reported as Indirect in nature on the Forms 4.
Source: Thomson Financial

NOTES

Prologue

1. Daniel Gross, "The Fiber Optic Bubble," *Milken Institute Review*, First Quarter 2002.
2. Ibid.
3. Ibid.
4. KMI Research data.

CHAPTER 1   Bernie's Bad Idea

1. Scott Waller, "Ebbers 'Thankful' for 17 Years," *Clarion-Ledger*, May 1, 2002.
2. D. Victor Hawkins, LDDS It Happened Right Here in Mississippi, *Jackson Journal of Business*, February 1998.
3. Ibid.
4. Kevin Maney, "Small-town Exec Strikes Big Deal: Down-to-Earth Style Is Secret to His Success," *USA Today*, August 28, 1996.
5. Rob Urban, "WorldCom Founder Still King of Range," *Bloomberg News*, August 13, 2002.
6. "Area Laments Ebbers Leaving Post," *Clarion-Ledger*, Editorial, May 1, 2002.
7. Jayne O'Donnell, "Ebbers Acts as if Nothing Is Amiss," *USA Today*, September 19, 2002.
8. Amy Barrett with Peter Elstrom, "Making WorldCom Live Up to Its Name," *BusinessWeek*, July 16, 1997.
9. Kara Swisher, "Anticipating the Internet; Good Timing, Good Deal-Making and Good Luck Turned Rick Adams' UUNet into a Star," *Washington Post*, May 6, 1996.
10. David Faber Interesting People mailing list, http://www.interesting-people.org/archives/interesting-people/200207/msg00047.html.
11. Ibid.

12. WorldCom Press release, "WorldCom Announces $300 Million Expansion of UUNet Network," February 19, 1997.

13. Daniel P. Dern, "Intranet Visionaries, Part II: An Interview with John Sidgmore," *Telecommunications*, August 1997.

14. Henry Goldblatt and Nelson D. Schwartz, "Telecom in Play," *Fortune*, 83.

15. Brian Taptich, "Brave New WorldCom," *Red Herring*, November 1997.

16. WorldCom 1997 Annual Report.

17. David Staples, "A Telecom Prophet's Fall from Grace," *Edmonton Journal*, July 28, 2002.

18. Sarah Schafer, "For 2 Companies, Playful Co-existence," *Washington Post*, June 8, 2000.

19. Ibid.

20. Blaise Zerega, "The Next Ma Bell," *Red Herring*, May 1999.

21. First Interim Report by Bankruptcy Court Examiner Richard Thornburg, November 4, 2002 at http://news.corporate.findlaw.com/hdocs/docs/worldcom/thornburgh1strpt.pdf.

22. First Interim Report by Bankruptcy Court Examiner Richard Thornburg, "Reserves and Their Role in Company's Financial Statements," November 4, 2002, 106.

23. First Interim Report by Bankruptcy Court Examiner Richard Thornburg, "WorldCom's Relationships with SSB and Jack Grubman," November 4, 2002, 81.

24. Jessica Sommar, "Here Comes the Bribe," *New York Post*, August 30, 2002.

25. First Interim Report by Bankruptcy Court Examiner Richard Thornburg, "WorldCom's Relationships with SSB and Jack Grubman," November 4, 2002, 98.

26. CNBC Business Center with Ron Insana, MCI WorldCom CEO Bernard Ebbers and Sprint CEO William Esrey discuss their companies' merger plans, October 5, 1999.

27. Goldberg August 2002 Study, "WorldCom Employee Stock Options," and SSB's Atlanta Brokers Group, 63, Footnote 1, Salomon Smith Barney report on WorldCom by Jack Grubman, May 24, 1999, 1.

28. Ibid., 69, Footnote 2, SSB report on WorldCom by Jack Grubman, July 12, 2000, 1.

29. Ibid., 75, Footnote 3, SSB reports on WorldCom by Jack Grubman, November 1, 2000, 1.

30. Andrew Backover, "Ebbers Linked to $4 Million in Farm Subsidies," *USA Today*, November 12, 2002.

31. First Interim Report by Bankruptcy Court Examiner Richard Thornburg, November 4, 2002, 5.

32. Jonathan Krim, "Fast and Loose at WorldCom," *Washington Post*, August 29, 2002.

33. Ibid.

34. Ibid.

CHAPTER 2   Rocky Mountain High

1. Lisa Levenson, "Qwest's Nacchio, Telecom Survivor, Faces Big Rivals, Slow Times," *Bloomberg Markets,* October 2001.

2. Carleen Hawn, "No Ordinary Joe," *Forbes*, August 7, 1999.

3. "Key AT&T Executive, Joseph P. Nacchio, to Join Qwest Communications as President and CEO," PR Newswire, December 22, 1996.

4. Linda Wommack, "The Richest Square Mile on Earth," *American Western Magazine*, March 2001, at http://www.readthewest.com/wommackMAR2001.html.

5. Kathleen Morris and Steven V. Brull, "Qwest's $7 Billion Man," *BusinessWeek*, December 8, 1997.

6. Brian O'Reilly, "Billionaire Next Door," *Fortune*, Sept., 6, 1999.

7. Jerd Smith, "The Man with the Cash," *Rocky Mountain News*, December 21, 1997.

8. Center for Responsive Politics, at www.OpenSecrets.org.

9. Jeff Smith, "Is Troubled Qwest Missing Anschutz's Golden Touch?," *Rocky Mountain News*, April 27, 2002.

10. O'Reilly, "Billionaire Next Door."

11. Smith, "Troubled Qwest."

12. Ibid.

13. O'Reilly, "Billionaire Next Door."

14. "Anschutz Offers $50 for Rio Grande Shares," United Press International, October 1, 1984.

15. "Southern Pacific and Denver & Rio Grande Western Railroads Apply to ICC to Operate under the SP Banner—Company Press Release," *Business Wire* via Lexis-Nexis.

16. John Eckhouse, "Planting an Information Superhighway," *San Francisco Chronicle*, November 24, 1993.

17. Ibid.

18. Anita Raghavan, "For Salomon, Grubman Is the Big Telecom Rainmaker," *Wall Street Journal*, March 25, 1997.

19. Levenson, "Qwest's Nacchio."

20. Jerry Useem, "The New Entrepreneurial Elite," *Inc. Magazine*, December 1997.

21. Ibid.

22. Ibid.

23. John Markoff, "Microsoft Invests in Qwest," *New York Times*, December 15, 1998.

24. Hawn, "No Ordinary Joe."

25. Ibid.

26. "Qwest Communications Appoints Afshin Mohebbi President and Chief Operating Officer," *Business Wire*, May 4, 1999.

27. Ibid.

28. Kris Hudson, "Qwest's Quiet Mover and Shaker Mohebbi Is Determined Leader and Nacchio's Right-hand Man," *Denver Post*, June 10, 2001.

29. House Committee on Energy and Commerce, Subcommittee on Oversight and Investigations, W. J. "Billy" Tauzin, Chairman, *Capacity Swaps by Global Crossing and Qwest: Sham Transactions Designed to Boost Revenues?* October 1, 2001.

30. Ibid.

31. Christopher Byron, "Is Qwest Playing Accounting Games?" *Red Herring*, January 2002.

32. Chris Bryon, "Qwest Insiders Share Recipe for Revenue Growth," *Red Herring*, April 2000.

33. Kris Hudson, "After 2 Years of Concerns, Qwest Revised Accounting," *Denver Post*, September 27, 2002.

34. David Milstead, "Testimony Sends Qwest President into Spotlight," *Rocky Mountain News*, September 24, 2002.

35. House Committee on Energy and Commerce, *Capacity Swaps by Global Crossing and Qwest.*

36. Jeff Smith and Lou Kilzer, "Pocketing Millions: Qwest Annual Stock Holders Meeting," *Rocky Mountain News*, July 30, 2002.

CHAPTER 3   Once a Junkie, Always a Junkie

1. Thomas Easton, "The $20 Billion Crumb," *Forbes*, April 19, 1999.

2. James B. Stewart, *Den of Thieves* (New York: Simon & Schuster [Touchstone Books], 1992), 224.

3. Ibid.

4. Julie Creswell with Nomi Prins, "The Emperor of Greed," *Fortune*, June 24, 2002.

5. Easton, "The $20 Billion Crumb."

6. Ibid.

7. Chris Palmeri, "The Drexel Connection at Global Crossing," *Business Week*, March 11, 2002.

8. Ibid.

9. Ibid.

10. Easton, "The $20 Billion Crumb."

11. Creswell and Prins, "The Emperor of Greed."

12. House Committee on Energy and Commerce, Subcommittee on Oversight and Investigations, W. J. "Billy" Tauzin, Chairman, *Capacity Swaps by Global Crossing and Qwest: Sham Transactions Designed to Boost Revenues?* October 1, 2001.

13. Rob Guth, "Scheme for Mammoth Global Network Draws Carrier Attention," *Network World*, January 1998.

14. Jonathan Burke, "Submarine Attack," *Red Herring*, November 1997.

15. Ibid.

16. Creswell and Prins, "The Emperor of Greed."

17. House Committee on Energy and Commerce, *Capacity Swaps by Global Crossing and Qwest.*

18. "Global Crossing Has Fallen on Hard Times," Sat News Online, October 4, 2001.

19. Julie Creswell, "Global Flame Out," *Fortune*, December 2001.

20. "The Long Fall," *News Hour with Jim Lehrer*, Jeffrey Kaye of KCET

Los Angeles reports on the corporate collapse of Global Crossing, March 21, 2002.

21. House Committee on Energy and Commerce, *Capacity Swaps by Global Crossing and Qwest.*

22. Creswell, "Global Flame Out."

23. Christopher Palmeri and Arlene Weintraub, "As Global Crossing Sinks, Gary Winnick Stays Dry," *BusinessWeek*, October 22, 2001.

24. Karen Kaplan and Elizabeth Douglass, "Global's Unsinkable Capitan," *Los Angeles Times*, July 4, 2002.

25. Press Release, November 28, 2000; "Historical Landmark Property in Beverley Hills Officially Opens as Global Crossing Plaza," link at www.Cerrell.com.

26. Michael Scherer, "Mother Jones 400," *Mother Jones*, March 5, 2001.

27. Ibid.

28. Ibid.

29. Michael Scherer, "Global Crossing May Be an Upstart in the Wireless Business, But Its Top Three Executives Are Old Hands when It Comes to Campaign Contributions," *Mother Jones,* March 5, 2001.

30. Easton, "The $20 Billion Crumb."

31. Ibid.

32. "Ménage à Trois of Deceit," *Daily Enron*, May 21, 2002.

33. Ibid.

34. House Committee on Energy and Commerce, *Capacity Swaps by Global Crossing and Qwest.*

35. Ibid.

36. Elizabeth Douglass, "Winnick Involved in Deals, E-Mails Show," *Los Angeles Times*, October 1, 2002.

37. Ibid.

38. Dennis Berman, Julia Angwin, and Chip Cummins, "As the Bubble Neared Its End, Bogus Swaps Padded the Books," *Wall Street Journal*, December 23, 2002.

39. Elizabeth Douglass, "Global's Lost Connection," *Los Angeles Times*, October 14, 2001.

40. House Committee on Energy and Commerce, *Capacity Swaps by Global Crossing and Qwest*, September 24, 2002.

CHAPTER 4   Billionaire versus Billionaire

1. Bob Brown, "King of Qwest Holds Court," *Network World*, October 12, 1998.
2. Tim Greene, "Competitors in the Wrong Race," *Network World*, February 1, 1999.
3. Kevin Maney, "How Level 3 Worked Its Way to the Main Floor," *USA Today*, April 1, 1998.
4. Tony Mack with Carleen Hawn, "Bell Buster," *Forbes*, September 7, 1998.
5. Royce Holland recalled this story in an interview in November 2002.
6. Rebecca Blumenstein, "The Path to the U.S. Fiber Cable Glut," *Wall Street Journal*, June 17, 2001.
7. Ibid.
8. Rachael King, "Too Much Long Distance," *Fortune*, March 15, 1999.
9. Mack and Hawn, "Bell Buster."
10. Ibid.
11. Sara Barton, "Level 3 4th-Qtr Loss Almost Triple Revenue," *Bloomberg News*, January 26, 2001.
12. Ashley Brown, "Colorado Stocks Fall This Week, Left by Level 3's 38% Decline," *Bloomberg News*, April 6, 2001.
13. Andy Serwer, "The Inside Story of Level 3," *Fortune*, July 22, 2002.
14. Jeff Smith, "Some Still Hold Out Hope for Level 3," *Rocky Mountain News*, August 31, 2001.
15. Steve Jordon, "Gut-Wrenching Experience," *Omaha World-Herald*, May 13, 2001.
16. Olga Kharif, "Fiber Optics: Beware of Buffett's Bet," *BusinessWeek Online*, August 26, 2002.
17. Company Press Release, "Open Letter to Investors," July 19, 2001.
18. Rob Reuteman and Jeff Smith, "James Crowe's Still Bullish on Network's Prospects," *Rocky Mountain News*, August 25, 2001.
19. Creswell and Prins, "The Emperor of Greed."

CHAPTER 5   The Attack of the Clones

1. Kristen Hays, "Enron 'Tilted-E' Sign Goes for $44,000 at Auction," Associated Press, September 25, 2002.

2. Ibid.

3. Ibid.

4. Jeff Manning and Gail Kinsey Hill, "Portland Subsidiary Mirrors Enron's Rapid Rise, Fall," *The Oregonian*, December 16, 2001.

5. Robert Bryce, *Pipe Dreams: Greed, Ego, and the Death of Enron* (New York: Public Affairs, 2002), 145–147.

6. Ibid.

7. Ibid., 194–195.

8. Erick Schonfeld, "The Power Brokers," *Business 2.0.*, January 2001.

9. Ken Branson, "To Market, to Market," *Telephony*, February 2001.

10. Terry Sweeney, "Telepathic Power," *Red Herring*, July 2000.

11. Bryce, *Pipe Dreams,* 245–247.

12. Tom Fowler, "Broadband Unit Hype Didn't Match Reality, *The Houston Chronicle*, January 18, 2002.

13. "Our History," City of El Paso web site, www.elpaso.com/about/history.asp.

14. Floyd Norris, "Making Money on Fiber, the El Paso Way," *New York Times*, September 27, 2002.

15. El Paso Corporate Press Release, "El Paso Corporation Announces Highlights of February 5 Analyst Meeting, Including Telecommunications Strategy," *PR Newswire*, February 5, 2001.

16. *United States Securities Exchange Commission v. Andrew Fastow,* U.S. Southern District of Texas, Houston, October 2002.

17. David Bloom, "Blockbuster and Enron Blow Out," *Red Herring*, March 26, 2001.

18. "Blockbuster and Enron to Launch Entertainment On-Demand Service Via the Enron Intelligent Network," *Business Wire*, July 19, 2000.

19. Ibid.

20. Bryce, *Pipe Dreams,* 281–282.

21. Ibid.

22. Rebecca Smith, "Blockbuster Deal Shows Enron's Inclination to All-Show, Little-Substance Partnerships," *Wall Street Journal*, January 17, 2002.

23. Adam Lashinsky, "Enron's Own Dot-Com Bubble Finally Popped," The Street.com, July 13, 2001.

24. Ibid.

CHAPTER 6   Fresh Prince of Hot Air

1. Jerry Useem, "The New Entrepreneurial Elite," *Inc. Magazine*, December 1997.
2. Dan O'Shea, "Murder by Numbers," *Telephony*, October 8, 2001.
3. Ibid.
4. Mike Mills, "Cellular Visionaries Raj and Neera Singh Took Calculated Risks in Building a Company, an Industry, a Fortune," *Washington Post*, February 16, 1998.
5. Ibid.
6. Ibid.
7. Useem, "The New Entrepreneurial Elite."
8. Teligent S-1 filing, October 7, 1997.
9. O'Shea, "Murder by Numbers."
10. Teligent S-1 filing.
11. Ibid.
12. Matt Andrejczak, "Mandl Takes Teligent," *Washington Business Journal*, October 17, 1997.
13. O'Shea, "Murder by Numbers."
14. Teligent S-1 filing.
15. Kevin Maney, "Former Execs 'Walk All Over' AT&T," *USA Today*, September 24, 1998.
16. "Winstar Communications Shares to Trade On NASDAQ National Market," *Business Wire*, June 24, 1994.
17. "Winstar Tainted by CEO, Associates' Pay Plans," *Bloomberg News*, January 4, 1995.
18. Ibid.
19. "Robern Industries to Change Name to Winstar Communications Inc.; Name Change Reflects Current Activities," *Business Wire*, October 28, 1993.
20. Charles Platt, "The 38-Gigahertz Breakthrough," *Wired Magazine* 7:12 (December 1999).
21. "Winstar Communications Inc. Announces a Strategic Position in the Wireless Communications Industry," *Business Wire*, February 15, 1994.
22. "Winstar Tainted by CEO, Associates' Pay Plans," *Bloomberg News*, January 4, 1995.

23. "Winstar Board Sued by Shareholders for Alleged Corporate Waste," *Bloomberg News*, January 10, 1995.

24. "Winstar Communications Completes Private Placement of $225 Million of Long-term Notes Through Morgan Stanley," *Business Wire*, October 23, 1995.

25. *NASD Department of Enforcement v. Jack Grubman and Christine Gochuico,* Disciplinary Proceeding Number CAF020042.

26. Winstar Press Release, "$2 Billion Winstar/Lucent Strategic Agreement to Expand Winstar's Broadband Network," *PR Newswire*, October 25, 1998.

27. Winstar 10-Q and 10-K filings (10-K405 March 3 and 10, 2000).

28. Winstar 10-K405/A filing, May 1, 2000.

29. Ibid.

30. Fred Dawson, "Regulatory Goo Buys More Time," *Communications Engineering Design*, December 1997.

31. Teligent and Teledesic, File #DA 96-1481, Federal Communications Commission, "In the Matter of Freeze on the Filing of Applications for New Licenses, Amendments, and Modifications in the 18.8–19.3 GHz Frequency Band," www.fcc.gov/Bureaus/Wireless/Orders/1996/da961481.txt.

32. "Teligent Raises $800 Million for Wireless Network Buildout," *ATM News Digest*, July 7, 1998.

33. O'Shea, "Murder by Numbers."

34. Williams Press Release, "Winstar and Williams Communications Announce Major Agreements," December 17, 1998.

35. Winstar Press Release, "Winstar and Wam!Net Enter into Strategic Agreement," January 10, 2000.

36. Andrew Backover and Kevin Maney, "Global Crossing Puts Other Swaps under Scrutiny," *USA Today*, February 28, 2002.

37. Julie Creswell, "Inside a Wireless Nightmare," *Fortune*, July 9, 2001.

38. O'Shea, "Murder by Numbers."

39. Creswell, "Inside a Wireless Nightmare."

40. O'Shea, "Murder by Numbers."

41. From the Editor, "How the Mighty Have Fallen," *Broadband Week*, November/December 2000.

42. *NASD Department of Enforcement v. Grubman and Gochuico.*

43. Creswell, "Inside a Wireless Nightmare."

CHAPTER 7   Nobody@Home

1. "AT&T Acquires TCI in $68 Billion Mega Deal," *CableDataCom News*, July 1998.
2. Shawn Tully, "The IPO Boom," *Fortune*, May 27, 1996.
3. Marc Gunther, "The Cable Guys' Big Bet on the Net," *Fortune*, November 25, 1996.
4. Lucien Roberts, "The Race for More Bandwidth," *Wired*, January 1996.
5. Ibid.
6. Gunther, "The Cable Guys' Big Bet."
7. Roberts, "Race for More Bandwidth."
8. Ibid.
9. Jason Krause, "Still Excites," *Industry Standard*, February 2000.
10. Peter Elstrom, "Excite@Home: A Saga of Tears, Greed and Ego," *BusinessWeek*, December 17, 2001.
11. Gunther, "The Cable Guys' Big Bet."
12. Krause, "Still Excites."
13. John Heilemann, "The Networker," *New Yorker*, August 11, 1997.
14. Elstrom, "A Saga of Tears."
15. @Home 10-K filing, March 31, 1998.
16. Elstrom, "A Saga of Tears."
17. Ibid.
18. Saul Hansell, "A Hitch to Marital Web Bliss," *New York Times*, June 9, 1999.
19. Joanna Pearlstein, "AT&T Gets San Francisco Cable License to Itself," *Red Herring*, July 28, 1999.
20. Krause, "Still Excites."
21. Peter D. Henig, "Excite@Home Takes Some Body Blows," Red Herring Online, August 6, 1999.
22. Rachel Konrad, Corey Grice, and John Borland, "Was Excite@Home Marriage Doomed at the Altar?" *News.com*, August 21, 2001.
23. Excite@Home 10-Q filing, August 16, 1999.
24. Henig, "Excite@Home Takes Some Body Blows."
25. Elstrom, "A Saga of Tears."
26. Peter Elstrom, "Thomas Jermoluk: Excite@Home's Second Residence," *BusinessWeek*, December 6, 1999.

27. "Excite@Home Reports Fourth Quarter and Fiscal Year 1999 Results; Cable Modem Subscribers Hit 1.15 Million; Pro Forma Revenues Grew 76% to $129 Million; Profitability Achieved," January 20, 2000 press release.
28. Erika Brown, "Follow Your Bliss," *Forbes*, October 9, 2000.
29. Elstrom, "A Saga of Tears."

CHAPTER 8   Teddy Gets Taken to the Cleaners

1. Kenneth Gilpin, "William Little, 58, Buyout Firm Founder, Dies," *New York Times*, September 19, 2000.
2. Richard Morgan and David Carey, "Forstmann's $2B Citadel Acquisition Signals Revival," *The Daily Deal*, January 16, 2001.
3. Jan Hoffman, "A Latter-Day Warbucks, Helping Children," *New York Times*," April 27, 1999.
4. Diana B. Henriques, "Refilling Forstmann's War Chest," *New York Times*, December 15, 1994.
5. Hoffman, "A Latter-Day Warbucks."
6. "Clinton & GOP: Telecom Law Will Create Jobs," CNN, February 8, 1996.
7. Kenneth Brown, "Understanding the CLEC Crisis," white paper, The Alexis de Tocqueville Institute, May 28, 2001.
8. Andrew P. Madden, "Fierce Competitors," *Red Herring*, November 1997.
9. Mairin Burns, "Know When to Fold 'Em: Risking Lenders' Ire, Private Equity Firms Call it Quits," *Investment Dealers Digest*, October 7, 2002.
10. Debra Lau, "In the Beginning, There Was Telecommunications," *Venture Capital Journal*, March 2000.
11. Kevin Gray, "The Summer of Her Discontent," *New York*, September 20, 1999.
12. Dan Primack, "Round II: Connecticut Bites Forstmann Little," *Private Equity Week*, March 4, 2002.

CHAPTER 9   The House (of Cards) that Jack Built

1. Steven Rosenbush, Heather Timmons, Roger O. Crockett, Christopher Palmeri, and Charles Haddad, "Scandals in Corporate America: Inside the Telecom Game," *BusinessWeek*, August 5, 2002.

2. Anita Raghavan, "For Salomon, Grubman Is the Big Telecom Rainmaker," *Wall Street Journal*, March 25, 1997.
3. Peter Elstrom, "The Power Broker," *Business Week*, May 15, 2000.
4. Ibid.
5. Mark Langer, "The Siskel and Ebert of Telecom Investing," *New York Times*, February 4, 1996.
6. Kevin Maney, "Small-town Exec Strikes Big Deal; Down-to-earth Style Is Secret to His Success," *USA Today*, August 28, 1996.
7. Raghavan, "Grubman Is Telecom Rainmaker."
8. Ibid.
9. Ibid.
10. Amy Feldman and Joan Caplin, "What Would It Take to Be the Worst Analyst Ever?" *Money*, May 2002; additional reporting by Erica Garcia.
11. Nils Pratley, "Where Will the Blame Fall?" *Guardian*, June 27, 2002.
12. Ibid.
13. Faith Keenan, "Wall Street Conflicts Make Analysts' Calls Suspect," *Bloomberg News*, June 30, 2002.
14. Elstrom, "The Power Broker."
15. House Financial Services Committee, Capital-markets subcommittee, report by Richard H. Baker, chairman.
16. Laurie P. Cohen and Dennis K. Berman, "How Analyst Grubman Helped Call Shots at Global Crossing," *Wall Street Journal*, May 31, 2002.
17. Elstrom, "The Power Broker."
18. *State of New York and Eliot Spitzer v. Philip Anschutz, Bernie Ebbers, Clark McLeod, Stephen Garofalo and Joseph Nacchio*, Supreme Court of the State of New York, September 30, 2002.
19. Citibank documents submitted to the Committee on Financial Services, headed by U.S. Representative Michael Oxley.
20. *State of New York and Eliot Spitzer v. Philip Anschutz et al.*
21. Ibid.
22. Ibid.
23. Ibid.
24. Ibid.
25. Ibid.

26. Ibid.
27. Ibid.
28. Feldman, Caplin, and Garcia, "Worst Analyst Ever."
29. Lead Plaintiff Carl McCall, Comptroller of the State of New York, as Administrative Head of the New York State and Local Retirement Systems and as the Trustee of the New York State Common Retirement Fund, on behalf of purchasers and acquirers of all WorldCom publicly traded securities, File No. 02 Civ 3288, October 14, 2002, 97.
30. Jack Grubman testimony at House Committee on Financial Services: "Wrong Numbers: The Accounting Problems at WorldCom," July 8, 2002.
31. Ibid.
32. Ibid.

CHAPTER 10   Canadian Rhapsody

1. Ross Laver, "Nortel's Driving Force," *MacLean's*, August 2, 1999.
2. Shawn Young, "Nortel Networks Puts Pedal to the Metal," *USA Today*, August 18, 2000.
3. Nortel Networks annual report for 1999.
4. Jason Krause, "Blinded with Science," *Industry Standard*, March 26, 2001.
5. Christopher P. Locke and Michael Copeland, "Hello, Lucent? The Big Red Zero Tries to Reinvent Itself," *Red Herring*, December 19, 2000.
6. "McGinn Walks the Walk," RedHerring.com, June 29, 1999.
7. Locke and Copeland, "Hello, Lucent?"
8. Lucent Press Release, "Fidelity Holdings' IG2, Inc. Subsidiary Announces Strategic Agreement with Lucent Technologies," April 13, 2000.
9. Stephanie N. Mehta, "Lessons from the Lucent Debacle," *Fortune*, February 5, 2001.
10. Jason Krause, "Pulling Lucent Back from the Brink," *The Industry Standard.com*, February 9, 2001.

11. Krause, "Pulling Lucent Back from the Brink."

12. Tyler Hamilton and Robert Cribb, "John Roth Speaks Out," *Toronto Star*, December 8, 2001.

13. Remarks by Morton Bahr, President, Communications Workers of America, 2002 CWA Legislative-Political Conference, Washington, D.C., March 3, 2002.

14. Ibid.

15. Mary Jander, "Fat Cats Feast, Despite Layoffs," *Light Reading*, September 24, 2001.

16. Lucent's proxy statement for its 2002 Annual Meeting of Shareowners, filed on December 28, 2001.

17. Henry Schacht's Opening Remarks, Lucent memos, November 8–9, 2000, from www.Findlaw.com.

18. Mehta, "Lessons from the Lucent Debacle."

19. Robert Cribb and Tyler Hamilton, "How a Giant Fell to Earth," *Toronto Star*, December 8, 2001.

20. Lisa Chadderdon, "Nortel Switches Cities," *Fast Company*, August 1998.

21. Quentin Hardy, "Lighting Up Nortel," *Forbes*, August 21, 2000.

22. Sonia Chopra, "Ever on the Faster Track," Rediff.com, March 11, 2000.

23. Hardy, "Lighting Up Nortel."

24. Canadian Broadcasting Corp., "Nortel CEO Roth Made $100 Million US in 2000," March 14, 2001, at www.CBC.ca.

25. Om Malik, "Phoning for Help: Wall Street's Love Affair with Telecom Might Be on the Verge of a Breakup," *Red Herring*, December 19, 2000.

26. Hamilton and Cribb, "John Roth Speaks Out."

27. J. P. Donolon, "Optical Dis-illusion," *Chief Executive*, November 2000.

28. Ibid.

29. Nortel Press Release, "Nortel Networks Confirms Guidance for 2000 and 2001, Growth Expected at Rates Significantly Greater than Market," December 14, 2000.

30. *Carol Fraser v. Nortel Networks, John Roth, William Connor, Frank Dunn and Chahram Bolouri*, Ontario Court of Jutice, Court File No. 01-CV-206248.

CHAPTER 11 The Dan and Desh Show

1. Om Malik, "Tapping Sycamore," *Forbes.com*, July 21, 1999.
2. "Entrepreneur in Residence, Desh Deshpande," *Red Herring*, November 2000.
3. Derek Nunes, "Not Roses all the Way for Deshpande," *Silicon India*, April 2000.
4. Paul Judge, "The Savvy behind Sycamore," *BusinessWeek*, December 20, 1999.
5. Malik, "Tapping Sycamore."
6. Om Malik, "The Networked Matrix," *Forbes.com*, January 26, 2000.
7. Matthew Schifrin and Om Malik, "Amateur Hour on Wall Street," *Forbes*, January 25, 1999.
8. "Redback Networks More than Triples Following IPO," *Bloomberg News*, May 18, 1999.
9. Eric Moskowitz, "IPO Antics," *Red Herring*, May 2, 2001.
10. Ibid.
11. Daniel Briere and Christine Heckart in *Network World*, April 26, 1999.
12. BancBoston Robertson Stephens press release, "What's Next in Next Generation Networks?" August 5, 1999.
13. *SEC v. Paul Johnson*, United States District Court, Southern District of New York, www.sec.gov/litigation/complaints/comp 17922.htm.
14. SEC Press Release, "SEC Sues Former Robertson Stephens Inc. Research Analyst Paul Johnson for Issuing Fraudulent Research Reports; Robertson Stephens Consents to Pay $5 Million to Settle Related Administrative Proceedings," Litigation Release No. 17922, January 9, 2003.
15. Ibid.
16. Ibid.
17. Om Malik and Nicole Koffey, "Scream More for Sycamore," *Forbes Digital Tool*, October 14, 1999.
18. Om Malik and Nicole Koffey, "Sycamore's Scorching Start," Forbes.com, October 22, 1999.
19. R. Scott Raynovich, "Valuations Dominate the Conversation," Red Herring.com, November 4, 1999.

20. David Rynecki, "Networking Stocks: Sky-High but No Buy," *Fortune*, January 24, 2000.
21. Melanie Warner, "Wooing Potential Customers with Options," *Fortune*, July 2000.
22. Melanie Warner, "Friends and Family," *Fortune*, March 20, 2000.
23. Ibid.
24. Jason Krause, "Friends, Family and Customers," *Industry Standard*, May 15, 2000.
25. Ibid.
26. Press release, "Internet Traffic Tops 350,000 Terabytes per Month According to Market Research Firm RHK," January 18, 2000.
27. Ibid.
28. Press release, "Global Optical Transport Capital Expenditure Will Increase at Least 36% in 2001 According to RHK," October 30, 2000.
29. Vickers' Insiders Sales Data.
30. Matthew French, "Prescient Deshpande Helped Lay the Groundwork for Digital Data Streams," *Mass High Tech*, August 12, 2002.

CHAPTER 12   Just an Illusion

1. Ted Appel, "Cerent's Ascent," *Press Democrat*, December 12, 1999.
2. Ibid.
3. Tom Rindfleisch, "A Perspective on the Origin of Cisco Systems," at http://smi-web.stanford.edu/people/tcr/tcr-cisco.html.
4. Pete Carey, "A Start-up's True Tale: Often-told Story of Cisco's Launch Leaves Out the Drama, Intrigue," *San Jose Mercury News*, December 1, 2001.
5. Julie Pitta, "Long Distance Relationship," *Forbes*, March 16, 1992.
6. Andrew Kupfer, "The Real King of the Internet," *Fortune*, September 7, 1998.
7. Cisco 10-K filing, Cisco annual report, 1998.
8. Andy Reinhardt, "Meet Cisco's Mr. Internet," *BusinessWeek*, September 13, 1999.
9. Om Malik, "Vinod's Time," *Red Herring*, February 13, 2001.
10. Om Malik, "Fiber Raj," *Forbes.com*, November 29, 1999.
11. Om Malik, "Corning's CEO, Roger Ackerman, on How to Keep the Sparkle," *Red Herring*, November 2000.

12. "David Huber, president and CEO, Corvis Corp.," *Light Reading*, August 2000, at http://www.lightreading.com/document.asp?doc_id=1428&page_number=6.

13. Yuki Noguchi, "Ciena Paying $25 Million to Settle Suit," *Washington Post*, January 24, 2003.

14. Om Malik, "Vinod Khosla Predicts Optical Shakeout," *Red Herring*, November 2000.

15. Tom Stein, "VCs overflow with telecom funding," *Red Herring*, December 19, 2000.

16. Om Malik, "Phoning for Help: Wall Street's Love Affair with Telecom Might Be on the Verge of a Breakup," *Red Herring*, December 19, 2000.

17. Eric Nee, "Cisco—How It Aims to Keep Growing," *Fortune*, February 5, 2001.

18. James Surowiecki, "Cisco-Holics Anonymous," *New Yorker*, May 21, 2001.

CHAPTER 13   The Swami of the Broadband Boom

1. George Gilder, *Telecosm: How Infinite Bandwidth Will Revolutionize Our World* (New York: Simon & Schuster [Free Press], September 2000).

2. Kevin Kelly, "George Gilder: When Bandwidth Is Free," *Wired*, September 1993.

3. Paul Kedrosky, "Now that Bandwidth Is a Bust, George Gilder Is Moving On," *National Post*, August 25, 2001.

4. George Gilder, "Mike Milken and the Two Trillion Dollar Opportunity," *Forbes ASAP*, April 1, 1995.

5. Gary Rivlin, "Madness of King George," *Wired*, July 2002.

6. George Gilder, "Fiber Keeps Its Promise," *Forbes ASAP*, February 1, 1997.

7. Ibid.

8. Rob Walker, "The Gildercosm," *Slate*, September 11, 2000.

9. Rivlin, "Madness of King George."

10. Eric Nee, "JDS Is No Optical Illusion," *Fortune*, September 4, 2000.

11. Aaron Patrick, "An Adelaide Boy Who Grew Up to Be a Crash Winner," *Age*, September 22, 2002.
12. Rivlin, "The Madness of King George."
13. George Gilder and Bret Swanson, "Telecosm Explosion," *Forbes ASAP*, February 19, 2001.
14. Rivlin, "The Madness of King George."

INDEX